PRIDOLIAN AND EARLY GEDINNIAN AGE BRACHIOPODS FROM THE ROBERTS MOUNTAINS FORMATION OF CENTRAL NEVADA

BY

J. G. JOHNSON, A. J. BOUCOT, and M. A. MURPHY

UNIVERSITY OF CALIFORNIA PUBLICATIONS IN GEOLOGICAL SCIENCES
Volume 100

UNIVERSITY OF CALIFORNIA PRESS

PRIDOLIAN AND EARLY GEDINNIAN AGE BRACHIOPODS FROM THE ROBERTS MOUNTAINS FORMATION OF CENTRAL NEVADA

PRIDOLIAN AND EARLY GEDINNIAN AGE BRACHIOPODS FROM THE ROBERTS MOUNTAINS FORMATION OF CENTRAL NEVADA

BY

J. G. JOHNSON, A. J. BOUCOT, and M. A. MURPHY

UNIVERSITY OF CALIFORNIA PRESS
BERKELEY · LOS ANGELES · LONDON
1973

Volume 100

Approved for publication January 21, 1972

Issued April 13, 1973

UNIVERSITY OF CALIFORNIA PRESS
BERKELEY AND LOS ANGELES
CALIFORNIA

◇

UNIVERSITY OF CALIFORNIA PRESS, LTD.
LONDON, ENGLAND

ISBN: 0-520-09447-6
LIBRARY OF CONGRESS CATALOG CARD NUMBER: 72-75516

CONTENTS

PRIDOLIAN AND EARLY GEDINNIAN AGE BRACHIOPODS FROM THE ROBERTS MOUNTAINS FORMATION OF CENTRAL NEVADA

BY

J. G. JOHNSON, A. J. BOUCOT, and M. A. MURPHY

ABSTRACT

Two major brachiopod assemblages assignable to the Pridolian and Gedinnian stages occur in the upper part of the Roberts Mountains Formation on the north side of the Roberts Mountains. The Pridolian and Gedinnian age assemblages are recognized in measured section well above fossiliferous Ludlovian age horizons and are represented by many collections of silicified brachiopods. These comprise three faunas assigned to the Pridolian, of latest Silurian age, and three faunas assigned to the early Gedinnian, of Early Devonian age. Other stratigraphically useful fossils, including graptolites, conodonts, and trilobites are interbedded. Age assignments of all groups are consistent with placing the Siluro-Devonian boundary approximately 680–690 ft below the top of the formation.

New species include *Skenidioides robertsensis*, *Tyersella jubar*, *Dalejina subfrequens*, *Dicaelosia nitida*, *Schizophoria paraprima*, *Salopina submurifer*, *Morinorhynchus punctorostra*, *Mesodouvillina costatuloides*, *Ancillotoechia gutta*, *Reticulatrypa neutra*, *Atrypina prosimpsoni*, *Gracianella reflexa*, *G. cryptumbra*, *Tenellodermis matrix*, and *Metaplasia lenzi*. *Gypidula pelagica lux* is a new subspecies.

INTRODUCTION

THE ROBERTS MOUNTAINS FORMATION was proposed by C. W. Merriam (1940) to encompass a unit of well-bedded carbonates in central Nevada between the Ordovician Hanson Creek Formation and the Lone Mountain Dolomite. At the type section, near Pete Hanson Creek on the west flank of the Roberts Mountains (fig. 1), these formations are found in sequence, and for a score of years the lack of adequate maps prevented an understanding of their true stratigraphic relations. In 1960 Winterer and Murphy showed that nearly the whole of the Roberts Mountains Formation grades laterally into the Lone Mountain Dolomite and they presented evidence indicating a reef origin for the dolomite and a reef-flank, off-reef, and basin origin for the Roberts Mountains Formation. The Roberts Mountains Formation occurs throughout much of Nevada as a band adjacent to, and west of, Silurian dolomites and it is known now that the line between these lithofacies can be traced all around the western and northern margins of the North American continent (Berry and Boucot, 1970). It is the line that separates the outer pelagic, or graptolitic, facies from the shelly facies of the shelf, and since it is known to have remained remarkably stable in position throughout the Silurian, and parts of the Early Devonian and Ordovician as well, it is widely regarded as having represented the edge of the early Paleozoic continental shelf.

The Roberts Mountains Formation was thought to be wholly Silurian as late as 1963 (Merriam, 1963:38). Johnson (1965:368) examined brachiopods collected by Winterer and Murphy and pointed out that Devonian as well as Silurian horizons were encompassed by the Roberts Mountains Formation. The present writers joined forces in 1964 to collect in bulk from all fossiliferous horizons and to make known the Roberts Mountains brachiopods which are rich and diagnostic. To-

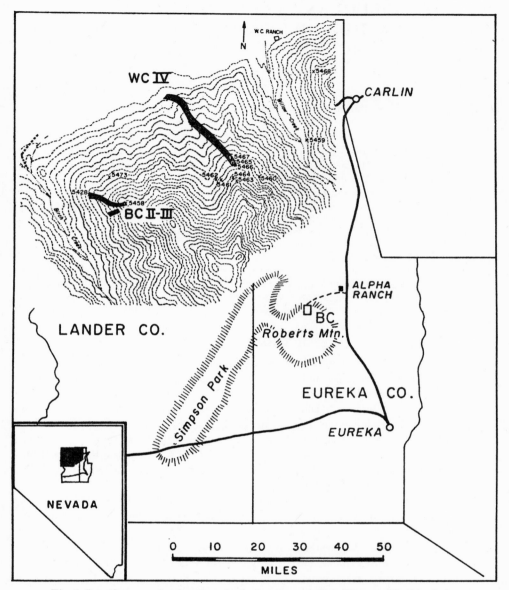

Fig. 1. Location map showing the principal sections and localities used in this study.

gether with the associated graptolites and conodonts they prove an age span of late Llandoverian to Gedinnian for beds in the area of the Roberts Mountains. A brief synopsis of early work, based largely on the brachiopods, has been given (Johnson, Boucot, and Murphy, 1967; Johnson and Boucot, 1968). The most recent work, on graptolites and conodonts by Murphy, W. B. N. Berry, and Gilbert Klapper, has confirmed most age assignments based on brachiopods and in many instances has allowed refinements in correlation which combine to make a faunal sequence for the late Llandoverian to Gedinnian interval that is unrivaled by any other yet reported in North America.

The present volume is the first of a series planned to deal with the brachiopods, graptolites, and conodonts. It treats the brachiopod faunas of Pridolian and Gedinnian age, the two youngest faunas of the Roberts Mountains Formation primarily, as developed on the north flank of the Roberts Mountains. A few collections from the Pete Hanson Creek area have also been included in the study. See figure 1.

Later volumes, by the present authors, will deal with the brachiopod faunas of Wenlockian and Ludlovian age. In addition, volumes dealing with the graptolites and conodonts of the Roberts Mountains Formation are in preparation by Berry, Klapper, and Murphy. These publications, on the paleontology of the fossils that are most useful for age dating, are to be followed by an account of the general geology and stratigraphy of the formation together with a coordinated account of the biostratigraphy of all elements of the fauna.

ACKNOWLEDGMENTS

We thank our colleagues W. B. N. Berry and Gilbert Klapper for aid and counsel during this project. Messrs. S. Finney, L. Price, J. Spencer, B. Wardlaw, and K. Wilson assisted in the field. The work was supported by National Science Foundation grants GA–17647 and GA–17455 to Oregon State University and GA–1035 and GA–12736 to the University of California, Riverside.

SEQUENCE OF FAUNAS

Beds of Pridolian and Gedinnian age crop out in the upper part of the section exposed at Birch Creek (Birch Creek section II–III) and occur in sequence above well-dated faunas of Ludlovian age. We have chosen to deal with the Pridolian-Gedinnian faunas together because of the continuity of the faunas in the section above 492 ft (UCR 5428), a horizon within the Pridolian, and because the question of the precise placement of the Siluro-Devonian boundary is a natural part of such a treatment.

At this time, however, we have not been able to recognize the Ludlovian-Pridolian boundary (assumed to be the Kopanina-Pridoli boundary) on the basis of the studies of brachiopods because the brachiopod faunas are poor between the lowest horizon treated here (UCR 5428) and the clearly Ludlovian faunas 426 ft below.

The Silurian faunas studied so far at Birch Creek and at a number of measured sections in the Pete Hanson Creek area indicate that, in the levels that can be confidently identified as Ludlovian, the pentamerids are ubiquitous and common in almost every collection. They drop out of the faunas without any indication of important environmental changes, as indicated by the rocks or by the associated taxa, as one ascends the section. The horizon at which the pentamerids completely disappear is apparently somewhere in the lower or middle Pridolian. This faunal change is referred to here as the Pentamerid Rule and leaves an interval of latest Silurian age that lacks Pentameridae, but which carries brachiopod faunas that are otherwise of Silurian type. This conclusion is based in part on unpublished evidence from conodont studies (Klapper, written communication, 1971) which suggest that pentamerids persist above levels that correlate with the highest

BIRCH CREEK
SECTION II-III

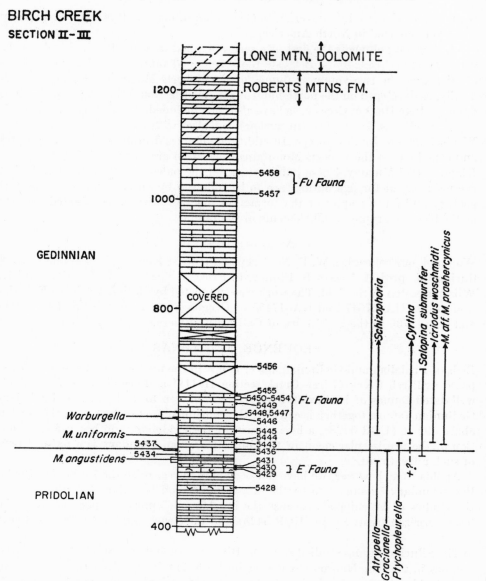

Fig. 2. Diagrammatic representation of Birch Creek section II–III showing the ranges and occurrences of some of the important taxa near the Silurian-Devonian boundary and the position of the localities in the Birch Creek II–III section. The horizontal line separating Pridolian and Gedinnian represents our correlation based on the entire fauna shown in the figure.

Kopanina and, therefore, range higher than the Ludlovian-Pridolian boundary. In any case, the horizon at which the pentamerids drop out in the Birch Creek section may be regionally significant as this change occurs in the nearby Pete Hanson Creek sections as well.

Pridolian.—At Birch Creek the fossiliferous section treated here begins with a collection from UCR 5428 (see fig. 2). This collection is similar to many of the confidently dated Ludlovian collections lower in the section and is relatively rich in individuals, but it lacks even a single specimen of pentamerid. The diagnostic Silurian elements are *Ptychopleurella* sp. E, *Atrypella* sp. E, *Gracianella cryptumbra*, *Delthyris* sp., and *Alaskospira?* sp. The complete brachiopod assemblage is as follows:

"Dolerorthis" sp.	*Gracianella cryptumbra* n. sp.
Ptychopleurella sp. E	*Nucleospira* sp.
Anastrophia sp.	*Meristina?* sp.
Gypidula sp.	*Howellella* sp.
Mesopholidostrophia? sp.	*Delthyris* sp.
Reticulatrypa neutra n. sp.	*Alaskospira?* sp.
Atrypella sp. E	

This fauna, though basically similar to other late Silurian collections from the Roberts Mountains, may be distinguished by the new species *Reticulatrypa neutra* and *Gracianella cryptumbra*, described from it and by the lack of any pentamerids.

From about 36 to 39 ft higher in the section above UCR 5428 there are three collections with a remarkable and distinctive brachiopod fauna which is here termed the "E fauna" for easy reference. The various brachiopod faunas in the region around the Roberts Mountains were given letter designations by Johnson and Boucot originally in 1968. The same scheme is used in this paper. The collections are from UCR 5429, UCR 5430, and UCR 5431. The faunal list from these localities includes the following species:

"Dolerorthis" sp.	rhynchonellids, indet.
Salopina sp. E	*Reticulatrypa?* sp.
dalmanellids, indet.	*Atrypella?* sp.
Gypidula sp.	*Gracianella reflexa* n. sp.
Leptaena sp. E	*G.* cf. *cryptumbra*
Morinorhynchus punctorostra n. sp.	*Protathyris* sp. E
Aesopomum? sp.	*Delthyris* sp.
Mesodouvillina costatuloides n. sp.	*Tenellodermis matrix* n. sp.
Lanceomyonia cf. *confinis* (Barrande)	*Cyrtina* sp.
Eoglossinotoechia? sp.	

Particularly distinctive in this fauna are the new species *Morinorhynchus punctorostra*, *Mesodouvillina costatuloides*, *Gracianella reflexa*, *Tenellodermis matrix*, and *Protathyris* sp. E. None of these five species occurs below and there is only a single occurrence of *Tenellodermis matrix* in higher beds (UCR 5434). It is worth a special note here to remark that *Cyrtina* sp. from this list is represented by a single articulated specimen. It is possible that this specimen entered the collection during laboratory processing because small, hollow specimens have been known to float in the wash water being used to deacidize the silicified residues. *Cyrtina* is an important constituent of the Gedinnian faunas and occurs commonly in many of the collections from those higher horizons. No other Silurian occurrence has ever been authenticated for the genus although certainly we must expect its ancestors to show a more complete record at some section not yet well studied.

In approximately the next 30 ft of section, succeeding the E fauna, three collections have been obtained which must be Silurian. These collections, UCR 5434, UCR 5436, and UCR 5437, include the following brachiopods:

Ptychopleurella sp. F
Tyersella? sp.
Dicaelosia nitida n. sp.
Salopina submurifer n. sp.
Salopina sp.
Gypidula sp.
Plectodonta? sp
orthotetaceans, indet.
Atrypa "reticularis" (Linnaeus)

Reticulatrypa neutra n. sp.
R. aff. *granulifera* (Barrande)
Dubaria megaeroides Johnson and Boucot
Atrypella sp. E
Gracianella cf. *cryptumbra* n. sp.
G. cf. *lissumbra* Johnson and Boucot
Meristina? sp.
Howellella sp.
Tenellodermis matrix n. sp.

This assemblage has a Pridolian-Gedinnian mixed character but we feel it must be assigned to the Silurian because of the presence of *Atrypella* and *Gracianella*. Also, connections with the older faunas must be noted in the form of the occurrence of *Reticulatrypa neutra* and *Tenellodermis matrix*. *Dubaria megaeroides*, which occurs in one of these collections and alone in a higher collection (UCR 5438), is associated with *Tenellodermis matrix* in the Tor Limestone and is also thought to be a Silurian brachiopod (Johnson and Boucot, 1970). With these taxa we note the first occurrence of *Dicaelosia nitida* and *Salopina submurifer*, species typical of the earliest Devonian fauna. Neither is abundant here, but their entry into the section is significant. The few scraps of *Ptychopleurella* assigned to *Ptychopleurella* sp. F also suggest affinity with Gedinnian faunas rather than those below. The remainder of the fauna is not diagnostic. The highest collection assigned to the Pridolian occurs at about 540 ft in Birch Creek section II–III. The collection, UCR 5438, falls slightly outside the measured section and its exact position in that section has not been determined.

Gedinnian.—There is fifteen foot interval above 540 ft which has not yielded any brachiopod collections, but at 555 ft in Birch Creek section II–III the basal Gedinnian fauna is first encountered. The Gedinnian assemblages are termed "F fauna." Three collections are grouped together for the purpose of this discussion. They are UCR 5443, UCR 5444, and UCR 5445. Together they include the following:

Ptychopleurella sp. F
Skenidioides robertsensis n. sp.
Protocortezorthis? sp.
Schizophoria sp.
Salopina submurifer n. sp.
Gypidula sp.
Leptaena sp. F?
Leptaena sp.
Aesopomum varistriatus Johnson
orthotetaceans, indet.
Strophonella sp.
Mesodouvillina cf. *costatuloides* n. sp.

rhynchonellids, indet.
Atrypa "reticularis" (Linnaeus)
Spirigerina marginaliformis Alekseeva
Dubaria sp.
Cryptatrypa sp.
Sibirispira? sp.
Nucleospira sp.
Meristina sp.
Howellella sp.
Undispirifer cf. *laeviplicatus* (Kozlowski)
Cyrtina sp.

To these, in all probability, may be added the float collection UCR 5439 and a collection labeled UCR 5442 that lost part of its field number but which is believed also to be a part of the UCR 5439 collection. Adding the taxa of these two to the three first mentioned would increase the list by the addition of *Lissatrypa* sp. This does not provide any new or unexpected associations because we have a collection, made by E. L. Winterer in which *Lissatrypa* sp. is abundant and which also includes the type specimens of *Aesopomum varistriatus*. In all likelihood the collections listed here as uncertain represent float from UCR 5443. At least it appears that *Ptychopleurella* sp. F, *Aesopomum varistriatus*, and *Lissatrypa* sp. represent a distinctive combination known nowhere but at this basal Gedinnian position in Nevada. The Devonian aspect is enhanced by the absence of Silurian genera such as *Gracianella* and *Atrypella*, which occur below, and by the first appearance of the typically, but not always, Devonian genus *Schizophoria*. *Skenidioides robertsensis* appears in the section for the first time in this assemblage and is common in most of the F fauna collections. The species *Spirigerina marginaliformis* occurs only in a single collection, but gives it a special distinctive appearance. *Undispirifer* cf. *laeviplicatus* appears here for the first time and is typical of higher collections. The same may be said for *Cyrtina* sp. except for the occurrence in UCR 5431, mentioned earlier (note the regular occurrence of this genus in collections from overlying beds; table 1). Also see figure 3.

About 24 ft above the basal Gedinnian collections the F fauna takes on a more stable character and several forms typical of the basal beds, that is *Ptychopleurella* sp. F., *Lissatrypa* sp., and *Spirigerina marginaliformis* do not occur again and the fauna for almost 100 ft, from UCR 5446 to UCR 5456 (see fig. 2 and table 1), is relatively easily characterized. It can be regarded as the lower F fauna. It includes the following species:

Skenidioides robertsensis n. sp.
Tyersella jubar n. sp.
Dalejina subfrequens n. sp.?
Dicaelosia nitida n. sp.
Schizophoria sp.
Salopina submurifer n. sp.
Gypidula sp.
Leptaenisca sp.
Aesopomum varistriatus? Johnson
orthotetaceans, indet.
Sphaerirhynchia sp.
Eoglossinotoechia? cf. *E. cacuminata* Havlíček
Atrypa "reticularis" (Linnaeus)
Dubaria sp.

Cryptatrypa angusta n. sp.
Cryptatrypa sp.
smooth atrypid, indet.
Atrypina prosimpsoni n. sp.
Rhynchospirina siemiradzkii Kozlowski
Nucleospira sp.
Meristina? sp.
Howellella sp.
Megakozlowskiella cf. *M.*? *inflectens* (Barrande)
Undispirifer cf. *laeviplicatus* (Kozlowski)
Metaplasia lenzi n. sp.
* *Cyrtina* sp.

Of these, forms with an asterisk are common and relatively characteristic of the fauna. The species *Cryptatrypa angusta, Atrypina prosimpsoni, Rhynchospirina siemiradzkii,* and *Metaplasia lenzi* occur for the first time in this fauna. Most of the species and individuals of the lower F fauna are small.

TABLE I

FAUNAL ASSEMBLAGES OF THE MOST IMPORTANT LOCALITIES USED IN THIS STUDY
(U.C.R. = University of California, Riverside, Department of Geological Sciences.
U.S.N.M. = United States National Museum)

SPECIES / LOCALITY (5428 U.C.R.; 10794* U.S.N.M.)	5428	5429	5430	5431	5432	5433	5434	5436	5437	5438	5439	5440	5441	5442	5443	5444	5445	5446	5447	5448	5449	5450	5451	5452	5453	5454	5455	5456	5457	5458	10794*	12865*	10795*	5459	5460	5461	5462	5463	5464	5465	5466	5467	5468	5471	5469	5470
AGE	PRIDOLI										?		LOWER GEDINNIAN																																	
FAUNA		E											F_L																Fu		F															
"Dolerorthis" spp.	●		●																																											
Ptychopleurella sp. E	●																																													
Ptychopleurella sp. F					?		?		●				●	●	●																															
Skenidioides robertsensis n. sp.														●	●	●	●		●	●	●	●	●	●	●	●	●			●					●		●			●				●		
Resserella elegantuloides (Kozlowski)																													●	●									?							
Tyersella jubar n. sp.																									●							●	●	●	●			●								
Tyersella? sp.								●																																						
Protocortezorthis? sp.																●																														
Dalejina subfrequens n. sp.																			?	?	?	?	?	?	?	?	?		?	●				?	?	?	?	?	●		?			●	●	?
Dicaelosia nitida n. sp.									●											●	●	●	●	●	●	●	●	●	●	●			●	●	●	●	●		●	●		●		●	●	
Schizophoria paraprima n. sp.																																				●	●							●	●	●
Schizophoria cf. *fragilis* Kozlowski																																		●	●			●								
Schizophoria sp.																	●		●	●	●	●		●		●	●	?		●											?	?				
Salopina submurifer n. sp.							●	●	●					●	●	●	●		●	●	●	●	●	●		●	●	●					●		●	●	●				●			●		
Salopina sp.								●			●	●																												●					●	●
Salopina sp. E			●																																											
Dalmanellids, indet.		●	●		●																																									
Anastrophia magnifica Kozlowski																														●				●											●	
Anastrophia sp.	●																																													
Gypidula pelagica lux n. subsp.																													●	●	●			●	●	●		●					●	●		
Gypidula sp.	●		●				●	●	●				●		●	●	●	●	●	●	●	●	●	●	●	●	●	●					●				●		●	●		●			●	●
Gypidula sp. F																																				●					●				●	●
Leptaenisca sp.																										●											●								●	
Leptaena sp. E		●	●																																											
Leptaena sp. F													?		?															●							?									
Leptaena sp.														●		●																				●					●			●		
Lepidoleptaena sp.																													●	●																
Plectodonta? sp.									●																																					
Morinorhynchus punctorostra n. sp.		●	●	●	●	●																																								
Iridistrophia cf. *umbella* (Barrande)																														●																
"Schuchertella" sp.																																				●	●				●			●		
Aesopomum varistriatus Johnson														?	●	?	●								?												●			?						
Aesopomum? sp.				●																																										
Orthotetaceans, indet.								●									●										●								●										●	
Leptostrophia sp.																																			?											
Strophonella sp.																●														●						●	●							●		
Mesodouvillina costatuloides n. sp.		●	●	●																																										
Mesodouvillina cf. *costatuloides* n. sp.														●			●																	●			●									
Mesopholidostrophia? sp.	●																																													
Stropheodontids, indet.						●																																							●	
Ancillotoechia gutta n. sp.																																		●	●											
Ancillotoechia? sp.																																									●					●
Sphaerirhynchia gibbosa (Nikiforova)																														●						●	●							●		
Sphaerirhynchia sp.																			?							●	●		●				?	●												
Hebetoechia? cf. *ornatrix* Havlíček																														●																
Hebetoechia? sp.																													●	●																
Lanceomyonia cf. *confinis* (Barrande)			●																																											
Eoglossinotoechia? cf. *C. cacuminata* Havlíček			?															●																		●	●							●		
Rhynchonellids, indet.		●															●																												●	
Atrypa nieczlawiensis Kozlowski																													●	●				●	●		●							●		
Atrypa "reticularis" (Linnaeus)	●						●						●		●				●	●	●	●	●	●	●	●	●	●					●			●			●	●	●	●			●	●
Reticulatrypa neutra n. sp.	●								●																																					
Reticulatrypa aff. *granulifera* (Barrande)							●	●																																						
Reticulatrypa? sp.				●																																										
Spirigerina marginaliformis Alekseeva																	●																													
Dubaria megaeroides Johnson & Boucot								●		●																																				
Dubaria sp.																●	?					●															●							●		
Cryptatrypa angusta n. sp.																						●	●		●	●		●								●	●							●		
Cryptatrypa sp.																?	●				●			●	●		●													●				●	●	
Atrypella sp. E	●						●																																							
Atrypella? sp.		?	●	●																																										
Atrypina prosimpsoni n. sp.																				●		●	●	●		●	●		●	●					●	●	●			●		●		●		
Gracianella reflexa n. sp.		●	●	●	●	●																																								
Gracianella cryptumbra n. sp.	●																																													
Gracianella cf. *cryptumbra* n. sp.				●					●																																					
Gracianella cf. *lissumbra* Johnson & Boucot							●	●	●																																					
Sibirispira? sp.																●																														
Lissatrypa sp.													●	●																																
Rhynchospirina siemiradzkii Kozlowski																						●			●	●		●		●				●	●	●		●		●					●	
Nucleospira spp.	●															●	●	●		●	●	●						●		●	●					●	●	●		●	●	●	●		●	●	●
Protathyris sp. E			●																																											
Meristella cf. *wisniowskii* Kozlowski																													●	●						●									●	
Meristina spp.	?						?		?						?	●	?						?		?		?								?	●										
Howellella spp.	●					●			●							●	●		●	●		●	●	●	●	●	●		●	●		?								●					●	
Delthyris spp.	●		●																																											
Megakozlowskiella cf. *M? inflectens* (Barrande)																											●														●				●	
Undispirifer cf. *laeviplicatus* (Kozlowski)															●						●														●	●	●			●						
Tenellodermis matrix n. sp.		●	●	●			●																																							
Metaplasia lenzi n. sp.																		●		●		●						●		●					●	●	●			●				●	●	
Alaskospira? sp.	●																																													
Cyrtina sp.				●													●	●	●		●	●	●	●	●	●	●							●	●	●	●			●				●	●	

Above the horizon of UCR 5456 there is a sequence of rocks, approximately 300 ft thick, that has yielded no brachiopods, but at a horizon about 1,012 ft (UCR 5457) and higher at about 1,050 ft (UCR 5458) the Gedinnian fauna takes on a new character dominated by brachiopods with medium- and large-sized shells. The following taxa were recognized:

Skenidioides robertsensis n. sp.
Resserella elegantuloides (Kozlowski)
Dalejina subfrequens n. sp.
Schizophoria sp.
Anastrophia magnifica Kozlowski
Gypidula pelagica lux n. subsp.
Leptaena sp. F
Lepidoleptaena sp.
Iridistrophia cf. *umbella* (Barrande)
Strophonella sp.
Sphaerirhynchia gibbosa (Nikiforova)

Sphaerirhynchia sp.
Hebetoechia? cf. *ornatrix* Havlíček
Hebetoechia? sp.
Atrypa nieczlawiensis Kozlowski
Atrypina prosimpsoni n. sp.
Rhynchospirina siemiradzkii Kozlowski
Nucleospira sp.
Meristella cf. *wisniowskii* Kozlowski
Howellella sp.
Metaplasia lenzi n. sp.

Fig. 3. Diagrammatic representation of Willow Creek sections IV and V showing the stratigraphic relationships of some localities used in this study. The location of these sections is plotted on figure 1 as WC IV.

The upper F fauna differs from the lower F fauna by the addition of *Resserella elegantuloides, Anastrophia magnifica, Gypidula pelagica lux, Iridistrophia* cf. *I. umbella, Sphaerirhynchia gibbosa,* and *Atrypa nieczlawiensis* and in the absence of *Salopina submurifer.*

Above the upper F fauna the change to Lone Mountain Dolomite occurs in less than 100 ft and no more fossiliferous horizons were observed. Higher pre-Nevada Group fossiliferous horizons are not found in the Roberts Mountains, but these have been identified at Coal Canyon, about 11 mi to the northwest, in the northern part of the Simpson Park Range (Johnson, 1970). Apparently, the next highest beds belong to a still poorly known brachiopod fauna that approximately corresponds to the graptolite zone of *Monograptus praehercynicus* which is in turn succeeded by the *Quadrithyris* Zone and younger zones monographed by Johnson (1970).

AGE AND CORRELATION

The lowest brachiopod faunule considered here (UCR 5428) lies about 400 ft, in measured section, above early Ludlovian age beds yielding brachiopods representative of the "C fauna" (Boucot and Johnson, 1968) of the Roberts Mountains Formation, so, by stratigraphic position, it must lie relatively high in the Silurian. Unfortunately there are no diagnostic graptolite collections from near the horizon of UCR 5428, but several collections of apparently post-Kopanina age conodonts have been obtained from lower levels (Klapper, written communications, 1971). This suggests a Pridolian age for UCR 5428. The pre-Gedinnian age of this locality is unimpeachable, judging from the whole assemblage which has a Silurian aspect (e.g., *Ptychopleurella, Atrypella, Gracianella, Delthyris,* and *Alaskospira?*). There are no brachiopods at all suggestive of the Devonian. As has been pointed out in an earlier section, however, correlation of these beds with either the late Ludlovian or the Pridolian has not yet been possible and we have assigned the pre-Gedinnian beds, above the highest pentamerids, to the Pridolian as a working hypothesis. The Pentamerid Rule seems useful for the Great Basin and perhaps for western and Arctic North America. It may also have meaning in the Greben Formation of Vaigatsch (Nikiforova, 1970). Common pentamerids occur only in the pre-Skala beds in Podolia (Kozlowski, 1929). In other regions of the USSR (viz., the Urals and Central Asia) it seems possible that pentamerids occur well into strata of Pridolian age, but detailed stratigraphic work on the faunas in those regions, based on collections obtained in measure sections, would be desirable.

The E fauna presents a similar problem. It includes an easily recognized assemblage, but one that is seemingly endemic at the species level and therefore makes long-range correlation, on the basis of the brachiopods, presently impossible. Late Silurian age and Pridolian assignment again is based largely on position in the sequence, lack of pentamerids, and lack of Gedinnian elements. Although a lone specimen of *Cyrtina* occurs with the underlying assemblage, the abundance of *Gracianella* attests to the Silurian age. In western North America so does the presence of *Delthyris,* although that genus certainly ranges into the Lower Devonian in eastern North America (Boucot and Johnson, 1967). *Monograptus angustidens* occurs above the E fauna beds at three slightly higher horizons (516, 522, and 526 ft according to Berry, written communication, 1971), emphasizing the nearness of the E fauna to fossils that can be correlated directly with beds of Pridolian age in the Old World. The highest three collections that are assigned to the Pridolian closely succeed the beds with *Monograptus angustidens*. In this highest brachiopod fauna, assigned to the Silurian, significant entries are *Dicaelosia nitida* and *Salopina submurifer*. These are typical of the Gedinnian fauna which is not only distinct in the local section but which is characterized by the final extinction of the Silurian genera and by the entry of Devonian genera and of species that can be correlated with the Old World Gedinnian.

The lowest collection of the basal Gedinnian fauna (UCR 5443) is in a limestone ledge at 555 ft in Birch Creek section II–III, immediately below a horizon at 557 ft with *Monograptus* aff. *praehercynicus* (Berry, Jaeger, and Murphy, 1971). *Monograptus uniformis* occurs first at 567 ft. These confirm an Early Gedinnian age assignment based on direct correlation with the faunal sequence in the Old World. In addition, the Early Gedinnian conodont, *Icriodus woschmidti*, was recognized 4 ft above *M.* aff. *praehercynicus*. The collection with *I. woschmidti* includes *Spathognathodus remscheidensis* of a variety known in the Lower Gedinnian of Europe (Klapper *in* Johnson, Boucot, and Murphy, 1967; and Klapper *in* Berdan and others, 1969). Thus, the brachiopods, conodonts, and graptolites are all in close agreement for a basal Gedinnian assignment of beds in the 10–12 ft strata centering around 560 ft in the Birch Creek sequence, and it appears that from the evidence available the boundary itself lies between 540 and 555 ft in that measured section.

A series of collections through a sequence of about 100 ft (approximately 600–700 ft in Birch Creek section II–III) yielded a fauna called lower F fauna. Forms that are distinctive of the basal Gedinnian fauna and those that are transitional with the Silurian, or are the final holdovers from the Silurian faunas, are absent from this fauna. Instead, it is typified by a large assemblage of small brachiopods, including some Old World Gedinnian species, or forms comparable to them (e.g., *Rhynchospirina siemiradzkii*, *Megakozlowskiella* cf. *M? inflectens*, and *Undispirifer* cf. *laeviplicatus*). *Schizophoria* and *Cyrtina* are common in this fauna and it is also noteworthy that the Gedinnian trilobite *Warburgella rugulosa* occurs in the basal beds of this fauna at 598–599 ft and at a horizon 10 ft higher (Alberti, Haas, and Ormiston, 1971).

Adequate collections were not obtained in the Birch Creek section through an interval of about 300 ft above the lower F fauna, but at two horizons, high in the section (1,012 and 1,050 ft), beds with a diagnostic fauna of medium- and large-sized brachiopods occur. Some large brachiopods appear in the upper F fauna which do not occur in older beds (e.g., *Resserella elegantuloides*, *Anastrophia magnifica*, *Gypidula pelagica lux* n. subsp., *Iridistrophia* cf. *umbella*, *Sphaerirhynchia gibbosa*, and *Atrypa nieczlawiensis*). All these forms occur in the lower Gedinnian Borszczow beds of Podolia (Kozlowski, 1929; Nikiforova, 1954). The other species of the fauna are largely those that had already entered the section as part of the lower F fauna.

The increased size of the elements of the fauna that were represented in the lower F fauna may be attributed to evolution, to being deposited closer to the life habitat, or to more favorable ecologic conditions in the habitat. The introduction of new elements into the fauna and the character of the sediments—coarser grain size, thicker bedding, and large-scale cross beds—suggest evolution and proximity to source are more likely. If the differences between the upper and lower F faunas are consistent in other areas, it may be possible to eliminate all the ecologic and habitat factors and to recognize two zones in this part of the stratigraphic column based on the evolution of the F fauna.

PROVINCIAL AFFINITIES

The Pridolian brachiopods of the Roberts Mountains Formation do exhibit provincial relations in contrast with much of the Silurian braciopod fauna of the globe which was cosmopolitan in the broadest sense (Boucot and Johnson, in press). This provinciality is manifested in the difference of the Roberts Mountains Formation faunas from contemporaneous brachiopod faunas of eastern North America, but there seems to be endemism in both areas. Wherever the brachiopod genera testify to definite provincial affinities, the Roberts Mountains Formation faunas point to Old World relations. This is especially exemplified by the great abundance of *Gracianella* and the common occurrence of *Atrypella*. The first is unknown in the Appalachian Province and the latter is uncommon. The genus *Dubaria*, which is abundant at some horizons, is also an Old World genus that is unknown in the Appalachian Province.

This Old World affinity, developed in the latest Silurian brachiopod faunas of Roberts Mountains, seems to herald conditions of provinciality already documented for the Devonian when Appalachian and Old World provinces were the principal provinces to which brachiopods conformed (Boucot, Johnson, and Talent, 1969; Johnson and Boucot, in press).

The affinities of the brachiopod genera of Gedinnian age seem to require similar conclusions. Genera like *Skenidioides, Salopina* of the *crassiformis-sub-murifer* type, *Dubaria, Cryptatrypa, Anastrophia* of the *magnifica* type, and *Gypidula pelagica* all testify to Old World connections because, with one exception, they are unknown in the Appalachian Lower Devonian. That one exception seems to be *Gypidula pelagica* which may be represented by the rare species *Gypidula angulata* Weller. The typical *Gypidula* of the Gedinnian of eastern North America is *Gypidula coeymanensis*. An exception to the rule of Old World affinity appears to be *Atrypina prosimpsoni* which has its closest ties with *Atrypina imbricata* of the Heldergergian. It seems convenient at this time to designate the Pridolian brachiopod faunas of the Great Basin as part of an Old World Province in line with the conclusions already in print pertaining to the Devonian (Boucot, Johnson, and Talent, 1969).

SYSTEMATIC PALEONTOLOGY
Phylum BRACHIOPODA
Order ORTHIDA
Suborder ORTHOIDEA

Discussion.—The suborder Orthoidea has a sparse representation in the Pridolian and Gedinnian of central Nevada. True *Dolerorthis* is a convexo-plane genus that differs from the common dolerorthid found in the Silurian of Nevada. "*Dolerorthis*" does not extend into Gedinnian age beds in the Roberts Mountains, but this is probably attributable to ecologic control because the genus does reappear in the *Quadrithyris* Zone, representing the early Siegenian (Johnson, 1970). The typically Silurian genus *Ptychopleurella* is represented by an ordinary-looking species in the lowest fauna assigned to the Pridolian, but in higher beds the genus is represented by an unusually flattened species that extends up into basal Gedinnian age beds. *Ptychopleurella* is not verified in younger beds anywhere. The third genus of the Orthoidea is *Skenidioides*. It is a long-ranging and conservative genus with a poor distribution locally in the Pridolian, but it is a common member of the Gedinnian faunas in the Roberts Mountains.

Superfamily ORTHACEA Woodward
Family HESPERORTHIDAE Schuchert and Cooper
Subfamily DOLERORTHINAE Öpik
Genus *Dolerorthis* Schuchert and Cooper

Discussion.—*Dolerorthis* should be restricted to shells that, like the type species, have brachial and pedicle valves of about equal depth, this being achieved by pronounced convexity of the brachial valve and by a nearly flat pedicle valve whose depth owes to the development of the palintrope. Specimens referred to here as *"Dolerorthis"* are plano-convex. In the brachiopod Treatise (Williams and Wright *in* Williams, 1965:H318) this concept is included in *Schizoramma*, but *Schizoramma** has a unique feature, a pair of lateral ridges adjacent to the central ridge of the cardinal process in the notothyrial cavity. These ridges distinguish it from most morphologically similar species.

"Dolerorthis" spp.
(Pl. 1, figs. 9–13)

Material.—Ten silicified specimens from two localities.

Exterior.—The valves are transversely suboval to subquadrate in outline and plano-convex to unequally biconvex in lateral profile. The ventral interarea is prominent, triangular, only slightly curved, and apsacline, about halfway between the orthocline and catacline positions. It is cleft by a narrow, triangular, open delthyrium. The dorsal interarea is low, triangular, and anacline. Pedicle valves are only gently convex and tend to be subcarinate. Brachial valves are flat or gently convex and bear a faint to prominently developed median sulcus.

The ornament consists of rounded to subangular radial costae that increase in number anteriorly by bifurcation and by implantation. The radial costae are crossed by numerous fine concentric growth lines.

Interior of pedicle valve.—Hinge teeth are small and triangular, attached to very small dental plates; the latter bound a small transverse-oval muscle scar.

Interior of brachial valve.—Brachiophores are short and prismlike and diverge at a moderate angle, slightly less than 90°. They bound a slightly elevated notothyrial cavity with a median, bladelike cardinal process. The notothyrial cavity is extended medially on some specimens by a low rounded ridge or a myophragm. The interior of the brachial valve, and pedicle valve as well, is crenulated internally, strongest peripherally.

Discussion.—There possibly are two species involved in the collections from

* Bassett (1970) has indicated the name *Schizoramma* should be replaced by Schizonema.

the two horizons noted below. The higher horizon contains only two pedicle valves and they are a little larger than those from the other collection. They seemingly differ in having finer radial costellae that bifurcate more evenly and do not develop any hint of parvicostellation. In addition, the dental plates of the specimens from the higher horizon are stronger.

Occurrence.—Pridolian and E faunas; UCR 5428 and UCR 5430 at Birch Creek section II–III.

Figured specimens.—USNM 171275–171278.

Subfamily GLYPTORTHINAE Schuchert and Cooper
Genus *Ptychopleurella* Schuchert and Cooper
Ptychopleurella sp. E
(Pl. 1, figs. 1–8)

Diagnosis.—Small, domed *Ptychopleurella;* concentric ornament strongly lamellose.

Material.—Twenty-six silicified specimens.

Discussion.—This species is near the typical Silurian *Ptychopleurella* of the *bouchardi* type with both valves rather deeply convex and with rounded cardinal extremities. The next higher *Ptychopleurella* sp. F is distinct, so *Ptychopleurella* sp. E, representing the highest typically Silurian species, is of interest.

Exterior.—The valves are subequally biconvex in lateral profile, or the brachial valve may be slightly deeper. The valves are transversely subquadrate in outline. The cardinal angles are slightly obtuse and may be rounded. Maximum width is anterior to midlength. The ventral interarea is triangular, flat, and apsacline. The ventral beak is inconspicuous. The delthyrium is triangular and open. The anterior margin tends to be nearly straight or is slightly reentrant medially owing to strong sulcation of the brachial valve and the absence of a corresponding fold on the pedicle valve.

The ornament consists of relatively small, low costae on the pedicle valve and relatively prominent costae separated by deep interspaces on the brachial valve. Concentric ornament on both valves is strongly lamellose in closely spaced increments.

Interior of pedicle valve.—The hinge teeth are minute and bladelike, with their long axes parallel to the hinge line, and they are supported by vestigial, ridgelike dental lamellae that define the ventral muscle scar laterally. The muscle scar is very short and simple, developed as a simple surface, without faceting. The greatest width of the muscle scar is transverse and it is relatively cramped into the apical area without a prominent anterior margin. The margins of the interior are variably crenulate, reflecting the degree of costation.

Interior of brachial valve.—The cardinalia consist of a pair of widely divergent tusklike brachiophores forming the inner edges of the sockets without any support of fulcral plates, plus a knoblike cardinal process that nearly fills the notothyrial cavity. Muscle scars are not impressed although there is a faint, low, rounded, ridgelike myophragm in the posterior half of the valve. The interior is variably crenulated, reflecting the degree of costation.

Occurrence.—Pridolian; UCR 5428 at Birch Creek section II–III.

Figured specimens.—USNM 171271–171274.

Ptychopleurella sp. F

(Pl. 10, figs. 14–18)

Diagnosis.—*Ptychopleurella* of moderate to large size for the genus with relatively flat pedicle valve and with concentric ornament poorly developed.

Material.—Approximately twenty-four silicified specimens from several localities.

Exterior.—The valves are subequally biconvex in lateral profile or the brachial valve may be slightly deeper. The pedicle valve is typically relatively broad and shallow. The cardinal angles tend to be nearly at right angles and the hinge line is the place of maximum width. The ventral interarea is low, long, and triangular, cleft by an open delthyrium. The brachial valve is slightly sulcate medially.

The ornament consists of a typical development of relatively low, subangular costae crossed by concentric growth lines; the latter, however, are not strongly developed.

Interior of pedicle valve.—Hinge teeth are not preserved. The muscle scars are confined apically to a transverse fan-shaped area that may or may not be bounded anteriorly by a slight step at the edge of the muscle scar. The shell substance is relatively thick in larger specimens and is only faintly crenulated near the margins.

Interior of brachial valve.—Cardinalia are typical for the genus, composed of a pair of short, tusklike brachiophores without fulcral plate support and with a simple knoblike cardinal process. A low, rounded myophragm is variably developed in the posterior one-third of the valve. The interior is faintly crenulate in the same fashion as the pedicle valve.

Comparison.—This species is easily distinguished from the typical Silurian *Ptychopleurella bouchardi* and related forms, including *Ptychopleurella* sp. E, by the broad, relatively flat character of the pedicle valve.

Occurrence.—Pridolian and F fauna; this species has been noted in UCR 5437, UCR 5441, UCR 5442, and UCR 5443 at Birch Creek section II–III. The occurrence at UCR 5443 is the one certain Gedinnian occurrence of the genus on a worldwide basis. *Ptychopleurella* sp., possibly assignable to *P.* sp. F because of size, occurs at UCR 5432 and UCR 5434 at Birch Creek section II–III.

Figured specimens.—USNM 171377–171379.

Family Skenididae Kozlowski
Genus *Skenidioides* Schuchert and Cooper

Discussion.—*Skenidioides* is widespread around the world in rocks of Ordovician and Silurian age, but is unknown in beds younger than Silurian in the eastern part of North America (Appalachian Province). In the Old World, however, it ranges well up into the Middle Devonian (Boucot and others, 1966). *Skenidioides* is easily distinguished from its Appalachian Province counterpart *Skenidium* in having a much less strongly pyramidal pedicle valve, a much deeper spondylium, by the presence of a well-developed dorsal interarea, and by the absence of a pair of triangular hinge plates.

Skenidioides robertsensis n. sp.

(Pl. 10, figs. 1–13)

Diagnosis.—*Skenidioides* of spiriferoid form and low cataline profile.

Material.—A total of 378 silicified specimens.

Exterior.—The shells are broadly transverse, of spiriferoid form, and unequally biconvex in lateral profile. The brachial valve is only gently convex, but the long palintrope of the pedicle valve makes it four or five times as deep as the brachial valve. The ventral interarea is broad, flat, triangular, and steeply apsacline to catacline. The dorsal interarea is well developed, flat, and anacline. The delthyrium is open and unmodified, except internally, by the presence of the spondylium. The hinge line is long and straight and is the place of maximum width. The cardinal angles are acute and slightly pointed. The flanks of the pedicle valve slope off laterally with very little curvature and there tends to be a gentle medial humping giving the valve a subcarinate transverse outline. The brachial valve is gently curved with convex flanks divided by a shallow sulcus.

Both valves bear strongly rounded radial costae, separated by narrow, well-defined, U-shaped interspaces. The costae increase in number anteriorly both by splitting and by intercalation; both modes occur on each valve. No concentric ornament was observed.

Interior of pedicle valve.—Hinge teeth are ill-defined, but their tracks are joined by the posterior edges of a relatively deep, U-shaped spondlyium whose base rests on a short ridge, septum, or medial buildup of shell material in the ventral apex. Deeply incised pallial trunks are developed as radial furrows lateral to the median ridge. The remainder of the interior is smooth except for indistinct peripheral crenulations.

Interior of brachial valve.—The sockets are only moderately divergent, being set laterally against the dorsal interarea and floored by shallow socket plates whose medial edges are joined to a moderately deep, U-shaped cruralium. The cruralium is bisected throughout its length by a thin, bladelike septum with a small knoblike cardinal process at the apex. The median septum is extended anteriorly beyond the edge of the cruralium to about two-thirds of the length of the valve. Oval adductor muscle impressions, without bounding ridges, are evident beyond the anterior edge of the cruralium. The interior is smooth except for the radial corrugations on thinner-shelled specimens, reflecting the costation of the shell, and except for peripheral radial crenulations.

Comparison.—*Skenidioides robertsensis* differs from the Ludlovian age species *S. henryhousensis* Amsden (1958:149, pl. 14) in being relatively more transverse, having less curved flanks and in having a steeper, catacline, ventral interarea. The Podolian Devonian species of *Skenidioides* called *Scenidium lewisi* by Kozlowski (1929, pl. 1) differs from the new species in its less transverse outline and more curved palintrope. Among the named Devonian species *Skenidioides polonicum* Gürich, as illustrated by Biernat (1959, pl. 1), is coarser ribbed, less transverse, and has an apsacline ventral interarea. The Devonian occurrences of *Skenidioides* were reviewed by Boucot and others (1966: 364, 365). Among these there may be a few additional named species, but none is known internally.

Occurrence.—F fauna; UCR 5460, 5462, and 5465, west of Willow Creek. Also UCR 5442, 5443, 5444, 5445, 5447, 5448, 5449, 5450, 5451, 5452, 5453, 5454, 5455, and 5458 at Birch Creek section II–III. Also UCR 5473 at Birch Creek, but not in a measured section.

Figured specimens.—USNM 171371–171376.

Suborder DALMANELLOIDEA

Discussion.—The two superfamilies Dalmanellacea and Enteletacea have a strong representation in the Roberts Mountains Formation. *Resserella* is represented only in the higher fauna of the Gedinnian by an advanced, less tumid species, compared with forms known in the Silurian. So far, in Nevada, no Ludlovian or Pridolian *Resserella* has been found, although it is common in beds of Wenlockian age. *Tyersella* is abundant in a few collections in the F fauna of Gedinnian age. Elsewhere, it is known from southeastern Australia and from New York. Silurian precursors are known, but the phylogeny has not been studied to date. As in the Siegenian and Emsian of Nevada *Dalejina* is almost ubiquitous in Gedinnian collections and *Dalejina subfrequens* is an important member of the upper F fauna. No Pridolian age occurrences are known to date from Nevada. *Dicaelosia* has a very similar distribution, being common in beds of Gedinnian age and almost unknown in beds of Pridolian age, although it makes it reappearance in the column just below the Siluro-Devonian boundary. The Gedinnian species *Dicaelosia nitida* is a short-lobed type. The lineage of long-lobed forms is not known to extend above beds of Ludlovian age. It has a probable descendant in *Dicaelosia* aff. *varica* of the lower Windmill Limestone (Johnson, 1970).

The superfamily Enteletacea is represented by *Schizophoria* and *Salopina;* both are abundant in the Gedinnian and *Schizophoria* first begins in the Gedinnian in the Roberts Mountains, although it is proved in beds as old as Pridolian elsewhere in western North America (Lenz, 1970). The dominant *Salopina* of the Gedinnian is *S. submurifer* which belongs to a distinct species group including *S. crassiformis* (Kozlowski). A species of *Salopina* of the ordinary (*lunata*) type occurs in less abundance in beds of Pridolian age.

Superfamily DALMANELLACEA Schuchert
Family DALMANELLIDAE Schuchert
Subfamily RESSERELLINAE Lazarev, 1970
Genus *Resserella* Bancroft
Resserella elegantuloides (Kozlowski)
(Pl. 14, figs. 1–21)

Dalmanella elegantuloides Kozlowski, 1929:63, text figs. 9A, 10, 11; pl. 2, figs. 1–16.
Parmorthis elegantuloides Nikiforova, 1954:48, pl. 2, figs. 3–6.
Resserella elegantuloides Walmsley and Boucot, 1971:514, pl. 98, fig. 8, pl. 99, fig. 1.

Discussion.—Two features of the valve morphology especially characterize *Resserella elegantuloides*. These are the rather ordinary depth and curvature of the pedicle valve and the development of a parvicostellate external ornament.

Material.—A total of 709 silicified specimens from two localities.

Exterior.—Pedicle valves are broadly suboval and brachial valves are subsemicircular to subquadrate in outline and the valves are unequally biconvex in lateral profile. The pedicle valve is relatively gently curved and only two or three times as deep as the brachial valve. The latter is gently convex posteriorly, but tends to flatten out because of its broad sulcus anteriorly. The hinge line is straight and the interareas are relatively short, equaling only slightly more than half the maximum width. The ventral interarea is short, low, triangular, and low apsacline to orthocline. It is gently curved and cleft medially by an open triangular delthyrium. The dorsal interarea is flat, low, triangular, and anacline. The ventral umbo is not particularly prominent and the valve has a nearly even curvature across its width although some specimens tend to be subcarinate. The cardinal angles are obtuse and broadly rounded. Maximum width is slightly posterior to midlength. Curvature from posterior to anterior only decreases slightly toward the anterior margin. The brachial valve develops a shallow sulcus posteriorly and the sulcus broadens and flares prominently in a short distance anteriorly so that the whole anterior portion of the valve is flattened out. The anterior margin is virtually rectimarginate.

The external ornament consists of a parvicostellate pattern of fine and very fine radial costellae disposed in a resserellid pattern of costellation medially (Walmsley, 1965).

Interior of pedicle valve.—The hinge teeth are blunt and triangular in outline with the long sides of the triangles parallel to the hinge line. They are supported by a very short, low, platelike or almost ridgelike dental lamellae which define small umbonal cavities. The muscle field, which lies medially, is triangular, unfaceted, and very short, blending at its anterior margin with the interior of the shell without an intervening ridge or step. There may or may not be a broad, low ridge or myophragm bisecting the muscle field anteriorly. The interior is smoth except for peripheral radial crenulations.

Interior of brachial valve.—The cardinal process is very small, without an elaborate knoblike myophore. It appears most commonly to consist of an extremely short, rounded shaft lying in the notothyrial cavity and commonly connected with a broad, low, ramplike myophragm. The brachiophores are long, triangular, and prismlike, diverging only slightly and extending only slightly anterior of the vertical. Their bases curve around sharply to form small, short sockets without fulcral plates. The muscle field is elongate, narrowing slightly anteriorly, without bounding ridges. The remainder of the interior is smooth except for peripheral radial crenulations.

Occurrence.—F fauna; UCR 5457 and UCR 5458 at Birch Creek section II–III. In addition, there are eleven specimens of *Resserella* sp. (possibly *R. elegantuloides*) in UCR 5464, west of Willow Creek.

Figured specimens.—USNM 171426–171432.

Subfamily ISORTHINAE Schuchert and Cooper

Genus *Tyersella* Philip

Discussion.—This interesting genus is an isorthid, morphologically distant from *Isorthis* s. s.

Tyersella is typified by its long, rounded, unfaceted ventral diductor tracks and the narrow ridge separating them, and commonly by the anteriorly convergent muscle bounding ridges that enclose the diductor tracks laterally. In the brachial valve the sockets are short cylindroidal slots impressed in the shell substance rather than being formed basally by fulcral plates. The dorsal muscle scars are also characteristic with the posterior adductors being relatively small and triangular, typically cramped into the space between the brachiophores, a pattern somewhat reminiscent of rhipidomellids. *Tyersella* is present in the Appalachian Province Helderbergian where it is represented by the species *Orthis perelegans* Hall. *Tyersella typica, T. perelegans,* and the new species described below are the only ones that the writers can assign to the genus with certainty at this time although *Isorthis tetragona* (Roemer) may belong here. There are also some Silurian species that are very suggestive of *Tyersella*.

Tyersella jubar n. sp.

(Pl. 15, figs. 1–23; pl. 16, figs. 1–15)

Diagnosis.—Gently biconvex *Tyersella* with narrow, faintly impressed, ventral diductor scars.

Material.—A total of 434 silicified specimens from five localities.

Exterior.—In outline the valves are transversely suboval, modified only by a short hinge line. In lateral profile the valves are unequally biconvex with the pedicle valve about half again as deep as the brachial valve. The ventral interarea is very low, narrowly triangular, and curved, lying in the apsacline, nearly orthocline, position. The dorsal interarea is narrow, low, bandlike to faintly triangular, and anacline. Both interareas are relatively narrow, equaling about half the maximum width which commonly lies slightly posterior to midlength. From midlength to the anterior the outline is evenly semicircular. Posteriorly, the cardinal angles are broadly rounded and obtuse. The profile, viewed along the axis, displays nearly flat or very gently convex flanks on the pedicle valve, rounded across the midregions so that a subcarinate condition is not developed. The brachial valve bears a very broad, low, indistinct sulcus, not clearly bounded but blending laterally with the flanks of the valve. The result is more an anteromedial flattening than a true sulcation. The commissure resulting tends to be very slightly arched ventrally.

The ornament consists of numerous, closely spaced, rounded radial costellae that increase in number anteriorly by bifurcation and implantation. On one very well-preserved specimen it appears that intercalation is the common mode of increase on the brachial valve and that splitting is the common mode on the pedicle valve. There is a slight tendency to parvicostellation in the midregions of some specimens and this tendency becomes accentuated toward the anterior of larger specimens. The initial pattern of costation appears to include a median costella on the pedicle valve and a median pair of costellae on the brachial valve commonly, that is, the isorthid pattern of Kemežys (1968:88). The radial disposition of the costellae in the midregions is strongly modified on the posterolateral portions of the shells, where the costellae are curved more and more strongly concave toward the hinge line in response to the shortness of the interarea. Concentric ornament is minimal, consisting of a few faint, broadly spaced growth lines on some but not all specimens.

Interior of pedicle valve.—The hinge teeth are short and bladelike to slightly triangular in cross section and diverge from each other and the hinge line at a moderate angle. Their medial sides are grooved by crural fossettes and they are supported basally by short, thick dental lamellae which are continuous anteriorly with low, narrow, ridgelike muscle-bounding ridges that contain the diductor scars laterally and which tend to converge slightly anteriorly. The apex is commonly filled by a small triangular plug of shell material forming a pedicle callist. The adductor scars are generally not discernible, although they appear to be outlined as an elongate pair medially inside the diductor scars of one specimen (pl. 15, fig. 19). The diductor scars are elongate and smoothly rounded, without faceting. They narrow slightly toward the anterior from the base of the hinge teeth and dental lamellae and are divided medially by a very narrow, low, ridge-like myophragm. The muscle scars commonly extend about one-third to two-fifths of the distance to the anterior commissure. The interior is smooth except for fine radial grooving peripherally.

Interior of brachial valve.—The sockets are formed as impressed cylindroidal slots in the posterior portion of the shell unbounded posterolaterally, but bounded medially by the brachiophore bases. The latter are continuous with the brachiophores, forming slightly curved, thick triangular plates that diverge at slightly more than 90° in most specimens, but in a few the angle is about 90° or slightly less. No tendency toward the development of fulcral plates exists in the thick-shelled, full-sized specimens, although in very small specimens it is clear that the sockets were formed with fulcral plates as their bases (pl. 16, figs. 8, 9). The proximal ends of the brachiophores are close together, leaving only a very small area of the notothyrial cavity and consequently restricting the size of the cardinal process to a thin rod or blade that is not continuous anteriorly with any structure. The adductor scars together form an inverted pyriform outline in which the posterior adductors are small and triangular and closely cramped-in between the brachiophore bases. They are separated from the anterior adductors by an indistinct line of demarcation. The anterior adductors are impressed in the shell without bounding ridges for the most part and are roughly equidimensional and trapezoidal in outline. The medial pair of pallial markings are visible on several specimens. They emanate from the low rounded myophragm where it divides the anterior adductors and they continue anteriorly in a subparallel fashion but diverge slightly as they approach the margin of the shell. The remainder of the interior is smooth except for peripheral radial grooving.

Comparison.—*Tyersella jubar* differs from *T. typica* in having less-impressed and possible slightly broader ventral diductor impressions, and in being transverse compared with the slightly enlongate outline of *T. typica. Tyersella jubar* differs from *T. perelegans* in having narrower ventral diductor scars that are typically more gently impressed and divided by a much fainter myophragm. The brachial valve of *T. jubar* is slightly less sulcate than the brachial valve of *T. perelegans* and the latter appears to be relatively more transverse. If *Orthis tetragona* is a *Tyersella* it can be distinguished from *T. jubar* in being larger, broader, and more deeply biconvex.

Occurrence.—F fauna; USNM 10795, UCR 5459, and UCR 5460 at Willow Creek. These collections include almost all the available specimens. A few occur in UCR 5463, west of Willow Creek, and in UCR 5453, Birch Creek section II–III. In addition, there is a single specimen assigned to *Tyersella?* sp. at UCR 5436 (Pridolian), Birch Creek section II–III.

Figured specimens.—USNM 171433–171450.

Family RHIPIDOMELLIDAE Schuchert
Subfamily RHIPIDOMELLINAE Schuchert
Genus *Dalejina* Havlíček
Dalejina subfrequens n. sp.
(Pl. 12, figs. 20–24; pl. 13, figs. 1–24)

Diagnosis.—Subcircular *Dalejina* with small and indistinct ventral muscle scar divided by a long, anteriorly extended median ridge.

Material.—A total of 513 specimens from the type locality and 194 specimens from other localities. All are silicified.

Exterior.—The shells vary in outline from subcircular to slightly transverse-oval and the valves are subequally biconvex in lateral profile. The ventral palintrope is low and curved. The short, low, triangular ventral interarea is poorly defined and considerably less than half the maximum width. It is curved and apsacline. The dorsal interarea is poorly defined, narrow, and orthocline. The ventral beak extends only a slight distance posteriorly beyond the dorsal beak. Both valves are deeply convex, but in the small specimens the ventral posterior is subcarinate and the dorsal posterior tends to be slightly indented at the midline. Anteriorly the brachial valve is well rounded and the pedicle valve displays a slight tendency toward medial flattening. Maximum width is near midlength. The anterior commissure is rectimarginate.

The ornament consists of subangular to rounded radial costellae that split or add to their numbers by intercalation anteriorly. The costellae on the midsector are approximately radial, but those on the posterolateral flanks are arcuate with concave sides facing posteriorly. Pedicle valves display a basic medial pattern of a pair of costae at the midline which contrasts with that of the brachial valve that bears a median costella. This pattern is typical for the rhipidomellid brachiopods as noted by Kemežys (1968). The costellae are crossed by a few poorly developed growth lines that occur at irregular intervals.

Interior of pedicle valve.—The hinge teeth are bluntly triangular and supported by short, divergent, platelike dental lamellae. These generally enclose less than the posterior half of the muscle field which is impressed rather than defined by muscle bounding ridges. The diductor impressions are bilobed, indistinct, and simple, divided by a long, narrow ridge that extends anteriorly well beyond the margin of the diductor impressions. The available specimens do not show adductor impressions. The interior is smooth except for peripheral radial crenulations that consist of grooves separating the relatively flat-topped ridges which themselves bear minor radial grooves.

Interior of brachial valve.—The cardinalia consist of a stubby cardinal process and brachiophores whose bases curve to form sockets diverging anterolaterally at the posterolateral valve margins without any development of fulcral plates. The cardinal process, of rhomboidal outline, is situated in the narrow notothyrial cavity between closely adjacent proximal edges of the brachiophore bases and has a very short shaft. The base of the shaft lies on a low, rounded myophragm that bisects the dorsal adductor scars. These are indistinctly impressed, quadripartite impressions with the posterior pair narrower than the anterior pair. The latter fade almost completely into the dorsal interior so that its outline is very poorly defined, if at all. The remainder of the interior is smooth except for peripheral crenulations similar to those of the pedicle valve.

Comparison.—The number of named species of *Dalejina* is relatively large as the lists of Boucot and Amsden (1958:167) and of Boucot, Johnson, and Walmsley (1965:337, 338) attest. Nevertheless, *Dalejina subfrequens* can be immediately distinguished from those that are well known internally by the presence of its small, narrow, ill-defined ventral muscle scar and the long median ridge that divides the ventral interior. Only *Dalejina frequens* (Kozlowski, 1929, pl. 3) is relatively close to the new species which is named to draw attention to this morphologic similarity. Both have relatively small ventral muscle scars and long median ridges dividing the ventral interior, for example, see the two ventral interiors illustrated by Nikiforova (1954, pl. 3, figs. 8, 9). The principal difference between these similar species (which might have been regarded as geographic subspecies) seems to be in the slightly longer hinge line, slightly longer transverse aspect, and the anteriorly flatter pedicle valve of *Dalejina frequens*.

Occurrence.—F fauna; the species is confidently recognized at UCR 5458 at Birch Creek section II–III (the type locality), UCR 5464 west of Willow Creek, UCR 5473, and in UCR 5469 at 1,670 ft, Pete Hanson Creek. Numerous other localities yield *Dalejina* sp. which probably belongs to *Dalejina subfrequens,* but small size of many of the specimens as well as poor preservation prevents positive identification.

Figured specimens.—USNM 171415–171425.

Family DICAELOSIIDAE Cloud
Genus *Dicaelosia* King
Dicaelosia nitida n. sp.
(Pl. 16, figs. 16–24; pl. 17, figs. 1–7)

?*Bilobites bilobus* Kozlowski, 1929:60, pl. 1, figs. 24–29; not Linnaeus.

Diagnosis.—Short-lobed, plano-convex, multicostellate with relatively broad interareas.

Material.—A total of 765 silicified specimens.

Exterior.—The outline of the shells is truncated-cordate and the valves are plano-convex in lateral profile. The pedicle valve is only moderately curved from posterior to anterior and the umbo is slightly flattened and is not prominent. The ventral beak is short and only slightly protuberant beyond the hinge line. The ventral interarea is flat, low, narrow, triangular, and apsacline, cleft by an open triangular delthyrium of about 30°. The dorsal interarea is well developed, flat, low, triangular, and anacline. The hinge line is equal to approximately half the maximum width which is attained well anterior to midlength. The cardinal angles are obtuse, but angular rather than rounded, giving the valves a shouldered appearance. Each valve bears a strong, U-shaped sulcus producing a strong emargination of the anterior commissure.

The ornament consists of closely spaced radial or subradial costellae of variable size which on some specimens tend to develop a slight degree of parvicostellation. There is no noticeably stronger median costella down the midline of the separate lobes and there is a tendency for obsolescence of the costellae medially on both valves. Concentric ornament consists of a few ill-defined, widely spaced growth lines.

Interior of pedicle valve.—The hinge teeth are very small, semicircular, plate-like lobes whose bases are parallel to the hinge line, but whose tips twist to diverge anterolaterally. Beneath the interarea dental lamellae are extremely small and poorly developed. There commonly are no bounding ridges defining the ventral muscle field and it is therefore virtually imperceptible. The interior is divided into separate lobes by the median elevation corresponding to the sulcus and this median elevation may or may not be longitudinally grooved to form rodlike ridges. The periphery is deeply grooved radially, forming a series of short lobes inside the anterior (the eminences and embayments of Wright, 1968).

Interior of brachial valve.—The cardinal process is a small knoblike protuber-ance nearly filling the notothyrial cavity. Sockets lie laterally, defined by the valve margin and by straight, divergent, platelike brachiophores which also serve as socket plates. Adductor scars are not impressed. The medial and peripheral areas are grooved forming lobes as in the pedicle valve.

Shell structure.—The valves are endopunctate.

Comparison.—*Dicaelosia nitida* n. sp. bears a close resemblance to *Dicaelosia oklahomensis* Amsden (1951, pl. 15; 1968, pl. 8, fig. 4). They are similar in degree of lobation, costellation, and in the internal structures. *Dicaelosia nitida*, however, is larger, more lenticular, and has a relatively broader hinge line than *D. okla-homensis*. Probably, the deeply curved and prominent umbo and the very narrow interareas of *D. oklahomensis*, contrasted with the same elements in *D. nitida*, most easily serve to distinguish these two species.

Inevitably, *Dicaelosia nitida* must be compared with *Dicaelosia biloba* (Lin-naeus). The lectotype and paralectotype have been illustrated by Wright (1968, pl. 1). These two specimens are comparable in size and in convexity to *D. nitida*, but appear to have relatively narrower, less prominent interareas, broader diver-gence of the separate lobes, and more widely spaced primary costellae, forming a rather marked parvicostellate ornament. Unfortunately, *Dicaelosia biloba* is not well known internally. Curiously, Wright (1968) included under the name *Dicae-losia biloba* a second form or perhaps even two that differ from the types of *Di-caelosia biloba*. These specimens (Wright, 1968, pl. 1, figs. 5, 6, 13–17, and pl. 2, figs. 1–10) should be regarded as *Dicaelosia* sp. They have much in common with *Dicaelosia nitida* n. sp. These unnamed specimens from Britain and Gotland have similar lobation and costellation to that of *D. nitida*, but the British specimens appear to have narrower interareas and more prominently outlined ventral di-ductors scars. The specimens from Gotland figured by Wright (1968, pl. 2) and also by Amsden (1968, pl. 8, fig. 3) under the name *Dicoelosia biloba* (*sic*) seem to belong to a species with a very deeply arched pedicle valve, but one that is other-wise very similar to *D. nitida*.

In connection with these observations it may be noted that some additional re-study of possibly misidentified species is still to be undertaken. The species name *Dicaelosia bilobella* Amsden (1968, pl. 8, fig. 2) compares much more closely with *Dicaelosia biloba* than does the species listed under that name by Amsden on the same plate. The principal difference, judging from specimens of *D. biloba* and *D. bilobella* which have been illustrated, seems to be one of size.

Amsden has very adequately pointed out the significance of the convexity of the brachial valve of species in the *Dicaelosia biloba* group (Amsden, 1968:33, text fig. 21; pl. 8, fig. 5). *Dicaelosia varica* (Conrad) is a late Gedinnian (New Scotland age) species with a convex brachial valve and represents an advance in the direction of convexity relative to all other species of the *Dicaelosia biloba* group. Previously it was uncertain what significance could be made of this change from a plano-convex to a biconvex shell. But the presence of *Dicaelosia nitida* as a common element in the Early Gedinnian of Nevada serves to closely mark the level at which dorsal convexity becomes prominent (i.e., between the lower and upper Gedinnian), and this seems to be confirmed by finding *Dicaelosia* of the *varica* type in the lower unnamed fauna of the Windmill Limestone at Coal Canyon (Johnson, 1970).

The final species in the phylogeny emanating from *Dicaelosia biloba* is *D. dimera* (Barrande, 1879, pl. 91). It also is biconvex and is notable for its gentle anterior emargination and for its widely spaced costellae. *Dicaelosia dimera* occurs in beds of late Early Devonian age, but neither it nor any other species has yet been authenticated in beds as young as Middle Devonian, although the brachiopod Treatise (Wright *in* Williams, 1965) asserts that this is so.

Occurrence.—F fauna; USNM 10795, and UCR 5459, 5460, 5461, 5462, 5464, 5465, and 5467, all at Willow Creek. Also UCR 5437, 5448, 5449, 5450, 5451, 5452, 5453, 5454, 5455, 5456, 5457, and 5458, at Birch Creek section II–III. Also UCR 5473 Birch Creek, and UCR 5469 at 1,670 ft, Pete Hanson Creek.

Figured specimens.—USNM 171451–171458.

Superfamily ENTELATACEA Waagen
Family SCHIZOPHORIIDAE Schuchert and LeVene
Subfamily SCHIZOPHORIINAE Schuchert and LeVene
Genus *Schizophoria* King
Schizophoria paraprima n. sp.
(Pl. 10, figs. 19–34; pl. 11, figs. 1–11)

Diagnosis.—Small, biconvex, transverse *Schizophoria* lacking fold and sulcus; ventral diductor scars separated by a broad, flat ridge or myophragm.

Material.—A total of 195 silicified specimens.

Exterior.—The shells are transversely oval in outline and unequally biconvex in lateral profile with the brachial valve a little deeper than the pedicle valve. The beaks of both valves are both short and stubby and do not project much beyond flat or slightly curved adjoining interareas. The interareas are equal to about half the width of the valves; the ventral interarea is apsacline and slightly curved. The dorsal interarea is anacline to orthocline and flat. The hinge line is short and straight but joins broadly curved, obtuse cardinal angles. Maximum width is near midlength. The anterior commissure of the valves is smoothly rounded without being deflected into a fold and sulcus although there is a very gentle tendency for arching of the anterior commissure toward the brachial valve. The delthyrium is triangular and open and encloses an angle of slightly less than 90°.

The external ornament consists of numerous, very fine, radial costellae that increase in number anteriorly.

Interior of pedicle valve.—The hinge teeth are small and rounded at their tips. They are set off from the supporting dental lamellae by poorly developed fossettes. Dental lamellae are platelike and subparallel or only slightly divergent, reaching about two-fifths of the distance to the anterior commissure. The diductor tracks are divergent, narrow lobes separated by a broad, flat, low ridge, the ridge being broader than the individual lobes of the diductor scars. The interior is not crenulated, except peripherally.

Interior of brachial valve.—The sockets are small and delicate and are restricted to posterior portions of the cardinalia; the brachiophores and brachiophore supporting plates extend further anterolaterally. The bases of the sockets are formed by small concave plates connecting the brachiophore structures with the valve margin, forming fulcral plates. The brachiophores are thickened triangular structures connecting basally with more platelike brachiophore-supporting plates. The distal edges of the brachiophore–brachiophore-supporting plate structures are anteriorly concave. The cardinal process is an apically situated, rodlike structure connected to a thin, bladelike supporting shaft. The brachiophore-supporting plates extend anterolaterally or anteriorly in a curving fashion with their sides convex outward and their distal edges at the lateral junction between the posterior and anterior adductor scars. On some larger specimens lower ridges continue anteriorly and partly encircle the anterior adductors which are invariably separated by a ridgelike myophragm. The interior is smooth except for marginal crenulations and some poorly impressed pallial markings.

Comparison.—Externally *Schizophoria paraprima* resembles *S. fragilis*, but internally these species are easily distinguished because of the broad, low myophragm separating the ventral diductor scars in *S. paraprima*. *Schizophoria parafragilis* Johnson is elongate rather than transverse and is thicker shelled. Internally it only variably shows a short ventral scar such as occurs in *S. paraprima*. No other named species which is known internally in any way resembles *Schizophoria paraprima* in its external characters because younger species of *Schizophoria* generally are larger, more unequally biconvex, or have a well-developed ventral sulcus. Lenz (1970) illustrated *Schizophoria* from Pridolian age beds of Yukon Territory which differs from *S. paraprima* in having a narrow ventral myophragm.

Occurrence.—F fauna; the type collection is from UCR 5462. Another very similar collection with *S. paraprima* is UCR 5461. Both are west of Willow Creek. Also UCR 5473 near Birch Creek. In the Pete Hanson Creek area *Schizophoria paraprima* occurs in UCR 5469 at 1,670 ft and in UCR 5470 at approximately 1,870 ft.

Figured specimens.—USNM 171380–171390.

Schizophoria cf. *fragilis* Kozlowski
(Pl. 11, figs. 12–14)

Schizophoria fragilis Kozlowski, 1929:79, text figs. 19, 20; pl. 3, figs. 1, 2.
Isorthis orbicularis Boucot *in* Boucot and others, 1960:5, pl. 1, figs. 13–20; pl. 2, figs. 1–7; not Sowerby, 1839.

Material.—Thirty-nine silicified specimens.

Discussion.—This small species of *Schizophoria* is characterized by a long, narrow, ridgelike myophragm between the ventral diductor scars (Kozlowski, 1929, text fig. 20). Some of the early *Schizophoria* species have been discussed elsewhere (Johnson and Talent, 1967*b*). In addition to the specimens from the Sutherland River Formation of the Canadian Arctic Archipelago which Boucot *in* Boucot, Martinsson, and others (1960) first assigned to *Isorthis orbicularis,* this species also occurs in Gedinnian age limestone in the Klamath Mountains of northern California (USNM 11163–11165). The available specimens are too poorly preserved for description since only the posterior portions of valves are preserved. A number of the pedicle valve interiors show the long ventral diductor scars separated by a narrow ridge.

This form occurs almost at the top of the Roberts Mountains Formation in the northern Roberts Mountains in the horizon with abundant *Gypidula pelagica.* The horizons that have yielded *Schizophoria paraprima* n. sp. are lower in the Gedinnian sequence.

Occurrence.—F fauna; UCR 5459, 5460, and 5463, in the vicinity of Willow Creek.

Figured specimens.—USNM 171392–171394.

Record of *Schizophoria* sp.

Occurrence.—F fauna; UCR 5445, 5447, 5448, 5449, 5450, 5452, 5454, 5455, and 5458 at Birch Creek section II–III.

Material.—Fifty-seven specimens.

Subfamily DRABOVIINAE Havlíček
Genus *Salopina* Boucot
Salopina submurifer n. sp.
(Pl. 11, figs. 15–23; pl. 12, figs. 1–19)

Salopina cf. *crassiformis* Johnson and Talent, 1967*a*, pl. 9, figs. 15–27.

Diagnosis.—Small, transverse *Salopina* with elongate dorsal adductor impressions bounded by subparallel ridges and bisected by a long median ridge.

Material.—A total of 2,634 silicified specimens.

Exterior.—The shells are transversely subquadrate in outline and unequally biconvex in lateral profile with a deeply convex pedicle valve, three to four times as deep as the brachial valve. The ventral palintrope is relatively long and curved, but the beak is short and pointed. The ventral interarea is low, triangular, slightly curved, and apsacline; it equals slightly more than three-quarters of the maximum width. The dorsal interarea is well defined, flat, ribbon-like, and anacline. The ventral interarea is cleft by an open, triangular delthyrium that encompasses an angle less than 45°. The cardinal angles are obtuse and rounded so that maximum width is near midlength of the pedicle valve or slightly posterior of midlength in some brachial valves. The pedicle valve is strongly arched in transverse profile and subcarinate. The brachial valve bears a relatively deep and anteriorly broadening sulcus which, together with the subcarinate pedicle valve, forms a strong but evenly rounded ventral deflection of the anterior commissure.

The ornament consists of numerous, well-defined, rounded, radial costellae that increase in number anteriorly by bifurcation and by implantation. The radial ornament is crossed at irregular intervals by a few poorly defined, or on some shells well-defined, concentric growth lines.

Interior of pedicle valve.—The hinge teeth are small and triangular and are supported by short, widely divergent, platelike dental lamellae which enclose the muscle field. The muscle field is short and restricted to an apical position with adductors and diductors undifferentiated and without a dividing myophragm. The anterior edge of the muscle field is defined by a low ridge or step. The remainder of the interior is smooth except for peripheral crenulations which may emanate from about midlength.

Interior of brachial valve.—The cardinal process is a small, simple protuberance or it may be bilobed. The brachiophores are prism-like projections of triangular cross section that are directed ventrally and slightly anteriorly. Thin, U-shaped, fulcral plates form the bases of sockets between the lateral edges of the brachiophores and the interarea. In addition, the brachiophores are attached to the floor of the valve by supporting plates. The latter are subparallel in cross section, but diverge slightly anteriorly. Their anterior ends may or may not be connected with long, subparallel muscle-bounding ridges that outline the dorsal adductor impressions. In specimens in which the brachiophore supporting plates do not continue directly with muscle-bounding ridges, they lie slightly within those ridges as is typical of the dorsal structure of many salopinids. The adductor impressions are not divided into posterior and anterior pairs in most specimens, but in some there is a slight reentrant of the muscle-bounding ridges about at midlength marking the division between the posterior and anterior adductors. The muscle field is bisected by a long, low, median ridge that commonly extends well beyond the anterior margin of the adductor muscle field and in some specimens all the way to the anterior margin. The interior is smooth except for peripheral crenulations.

Discussion.—It is believed that by increase in the height of the dorsal median ridge, this species gave way to the genus *Muriferella* Johnson and Talent, 1967a. *Muriferella* differs from *Salopina submurifer* essentially only in the presence of a low, bladelike median septum in the brachial valve.

Comparison.—Walmsley, Boucot, and Harper (1969) reviewed ten species of *Salopina*. Of these, *Salopina submurifer* only resembles *Salopina crassiformis* (Kozlowski) because both bear the long median ridge in the brachial valve and long subparallel muscle-bounding ridges outlining the adductor impressions. *Salopina submurifer* differs from *S. crassiformis* in being more transverse and in having a relatively flatter brachial valve. In addition, these species may differ in that the dorsal muscle-bounding ridges of *S. crassiformis* converge anteriorly rather than being subparallel (see Kozlowski, 1929:69, fig. 13A). On the other hand, Kozlowski illustrated only the dorsal interior of a single specimen so this difference could be attributable to variation.

Walmsley, Boucot, and Harper (1969, pl. 74, figs. 10–12; pl. 75, figs. 1–4) figured specimens of *Salopina* from the Pembroke Formation of eastern Maine as *Salopina robitaillensis*. These specimens more closely resemble *Salopina crassiformis* and *S. submurifer* than they do *Salopina robitaillensis* as exemplified by the subparallel ridges bounding the dorsal adductor impressions. The discrepancy may be accounted for by the fact that the typical specimens of *Salopina robitaillensis,* from the Robitaille Formation, are of early Ludlovian age while those from the Pembroke Formation are younger and are of Pridolian or early Gedinnian age.

Occurrence.—F fauna; USNM 10795 and UCR 5460, 5461, 5462, 5467, all at Willow Creek. Also UCR 5434, 5437, 5442, 5443, 5444, 5445, 5447, 5449, 5450, 5451, 5452, 5454, 5455, 5456, at Birch Creek section II–III. Also UCR 5473 away from the measured section.

Figured specimens.—USNM 171395–171414.

Salopina sp. E
(Pl. 1, figs. 14–21)

Salopina sp. Lenz, 1970, pl. 83, figs. 18–26.

Material.—Eight silicified specimens.

Discussion.—This is a medium- to large-sized, thin-shelled species of *Salopina* characterized by a subcarinate pedicle valve and very fine, wirelike costellae. The ventral interarea is narrow, short, flat, and steeply apsacline. The dorsal interarea is steeply anacline. The brachial valve bears a very broad, shallow sulcus. Internally, the pedicle valve is virtually featureless except for very short dental lamellae. The brachial valve has platelike brachiophores and a ridgelike cardinal process. A dorsal myophragm and adductor muscle-bounding ridges are virtually nonexistent.

The available specimens are not obviously close to any of the species described by Walmsley, Boucot, and Harper (1969), but the very fine wirelike type of costellae suggests *Salopina lunata* as does one of Lenz's figures (Lenz, 1970, pl. 83, fig. 25) of a dorsal interior. The ventral muscle scar of *Salopina lunata,* however, is very characteristic and is not approached by the ventral scar of any of the available specimens, possibly because they are so small and thin-shelled. Much more material is needed before an adequate characterization of this species can be made.

Occurrence.—E fauna; UCR 5430, Birch Creek section II–III.

Figured specimens.—USNM 171279–171282, 171591.

Order PENTAMERIDA
Suborder SYNTROPHIOIDEA

Discussion.—The syntrophid brachiopods have a meager representation in the Pridolian and Gedinnian age beds of Nevada. A small, poorly known species of *Anastrophia* occurs in Pridolian age beds and the wide-hinged *Anastrophia magnifica* occurs in the F fauna of Gedinnian age.

Superfamily CAMERELLACEA Hall and Clarke
Family CAMERELLIDAE Hall and Clarke
Genus *Anastrophia* Hall
Anastrophia magnifica Kozlowski
(Pl. 17, figs. 8–22)

Anastrophia magnifica Kozlowski, 1929:140, text fig. 42; pl. 4, figs. 14–16.
Anastrophia magnifica Khodalevich, 1951:18, pl. 4, figs. 9a–d.
Anastrophia magnifica Nikiforova, 1954:65, pl. 4, figs. 1a–d.

Anastrophia aff. *internascens* Rukavischnikova, 1961:45, pl. 2, figs. 1–7; not Hall, 1867.
Anastrophia magnifica Kulkov, 1963:18, pl. 1, figs. 3, 4.
Anastrophia cf. *magnifica* Johnson, 1970:92, pl. 11, figs. 1–8.

Discussion.—This important species of *Anastrophia* was described by Kozlowski from the Lower Gedinnian Borszczow beds of Podolia. The synonymy entries above, however, plus known occurrences in the *Gypidula* 1—*Davidsoniatrypa* unit, Yukon Territory, of Lenz (1967) and in beds of early Siegenian age of the Mandagery Park Formation near Manildra, New South Wales, show that it is well established in the Siegenian as well as in the Gedinnian. The species from the Mandagery Park Formation will be described as new by N. M. Savage because of peculiarities possibly attributable to geographic variation. In any event neither the species *Anastrophia magnifica* nor any other species of *Anastrophia* is known as high as the Emsian, so its presence high in the Lower Devonian, in some instances, will aid in separating Siegenian and Emsian when better evidence is lacking.

Anastrophia magnifica is characterized by its relatively long, straight hinge line and shield-shaped transverse outline. The specimens from Nevada, both from the *Quadrithyris* Zone and from Gedinnian equivalents being described here, are very close to the Podolian type in this regard, although the latter generally attain a larger size.

Material.—Sixty-three silicified specimens from three localities.

Exterior.—The valves are transversely suboval to shield-shaped in outline; the more shield-shaped ones are relatively quadrate with nearly straight anterior margins. In lateral profile the valves are unequally biconvex with the brachial valve the deeper of the two. Brachial valves are deeply and evenly curved from posterior to anterior while pedicle valves have a somewhat flattened umbo and flat, or even slightly concave, posterolateral flanks adjoining the cardinal extremities. The latter may be obtuse, at right angles, or may have slight auriculations, depending on the growth stage and variability encountered. Although the hinge line is straight for some distance there is no interarea and the ventral posterior is cleft by a triangular open delthyrium encompassing an angle of about 60°. The flattened ventral umbo gives way anteriorly to a broad, shallow median sulcus which projects dorsally as a tongue that is accommodated by a low, flat dorsal fold. In some specimens the anterior commissure is asymmetrical, not developing a true fold and sulcus.

The valves bear relatively numerous, strong, subangular costae separated by deep U-shaped interspaces. There commonly are five to seven costae on each flank of pedicle valves and there commonly are five costae in the ventral sulcus although this number is variable. Costae typically are simple, but some branching occurs anteriorly on some specimens near the parietal margins of the fold and sulcus. On some specimens the median ventral costa splits into a pair of costae on the umbo. Concentric ornament consists of a few poorly defined growth lines at irregular intervals; however, they become more prominent and more numerous anteriorly on some specimens. The costae form a crenulated margin whose points interlock at the commissure.

Interior of pedicle valve.—The hinge teeth are thickened and grooved triangular protuberances (pl. 17, fig. 10). Beneath the delthyrium and in the posterior part of the valve there is a Y-shaped rhomboidal spondylium supported by a short, low median septum that reaches no further than the anterior edge of the spondylium. The interior is radially ridged, reflecting the costation of the shell.

Interior of brachial valve.—The cardinalia consist of very small, shallow, delicate, anterolaterally diverging sockets and a pair of small inner plates inclined toward the midline. These are connected with long, subparallel or slightly converging, outer plates that meet the base of the valve along discrete subparallel tracks. Basolaterally divergent alae are present (pl. 17, fig. 22). The interior is radially ridged, reflecting the costation of the shell.

Occurrence.—F fauna; a single specimen from UCR 5459, Willow Creek; UCR 5458, Birch Creek section II–III; UCR 5469 at 1,670 ft, Pete Hanson Creek.

Figured specimens.—USNM 171459–171464.

Suborder PENTAMEROIDEA

Discussion.—As noted in previous discussions, brachiopods of the family Pentameridae have not been found to range up into beds as young as Pridolian age in Nevada. Instead, the interval Pridolian-Gedinnian is dominated by the family Gypidulidae, and in some collections by an abundance of the species *Gypidula pelagica.*

Superfamily PENTAMERACEA M'Coy
Family GYPIDULIDAE Schuchert and LeVene
Subfamily GYPIDULINAE Schuchert and LeVene
Genus *Gypidula* Hall
Gypidula pelagica (Barrande)

Pentamerus pelagicus Barrande, 1847:469, pl. 22, fig. 3.
Pentamerus pelagicus Barrande, 1879, pl. 22, figs. 2g, 3; pl. 23, figs. 1–15; ?pl. 94, case II; ?pl. 108, case III.
Sieberella cf. *galeata* Kozlowski, 1929:135, fig. 39, pl. 6, figs. 1–3; not Dalman.
Gypidula pelagica Nikiforova, 1937:27, pl. 4, figs. 15, 16.
?*Gypidula* cf. *galeata* Nikiforova, 1937:27, pl. 4, fig. 19; not Dalman.
?*Gypidula olga* Khodalevich, 1939:15, pl. 11, fig. 8; pl. 15, figs. 6–8.
?*Gypidula* sp. McLaren, Norris, and McGregor, 1962, pl. 8, figs. 1–3.
Gypidula pelagica Johnson, 1970:97, pl. 15, figs. 12–15.

Discussion.—*Gypidula pelagica* is an important fossil because it appears to be abundant, restricted to a relatively small stratigraphic interval, and is very widespread geographically. The species is characterized by a nearly smooth exterior that on some shells is modified by faint plication of the ventral fold, and by the prominent trapezoidal tonguelike extension of the dorsal sulcus. The Bohemian specimens develop a tendency to have the ventral fold divided by an indistinct median furrow which extends to the anterior commissure and gently indents the sulcal tongue at the midline. Almost all the specimens from Nevada are completely without plication and lack the median furrow of the pedicle valve, but a few of the larger specimens from the collection of unsilicified material show the long faint plications of the midregions which help in identifying the species. The majority of the specimens, however, are not plicate and these are identical in their morphology to specimens from a Gedinnian horizon at Cathedral Mountain of the Northwest Territories in collections made by A. C. Lenz. These seem closest to the specimens illustrated in figures 5, 7, and 8 of plate 23 of Barrande

(1879). The specimen illustrated by Kozlowski (1929, pl. 6, figs. 1–3) also belongs in this group although it is a little less elongate than Bohemian specimens that lack any form of plication.

Gypidula pelagica lux. n. subsp.

(Pl. 18, figs. 1–18; pl. 19, figs. 1–12)

Diagnosis.—A subspecies of *Gypidula pelagica* almost completely lacking in plications and with the curvature of the brachial valve strongly developed posteriorly, but much less so anteriorly.

Material.—A total of 1,175 specimens, all but twenty-eight of which are silicified.

Exterior.—The valves are broadly pyriform in outline and unequally biconvex in lateral profile. Both valves are very strongly curved, but the pedicle valve is arched over the dorsal umbo with an evenly decreasing amount of curvature toward the anterior. Brachial valves are very strongly curved in the posterior and slope away to a much more gradual curve from somewhere near midlength to the anterior. The ventral palintrope is short, curved, and apsacline or nearly orthocline in its inclination and the ventral beak is bluntly pointed and strongly curved, lying near to, but not touching, the dorsal umbo. The delthyrium is triangular and open, without any deltidial covering. The hinge line is relatively long, slightly curved, and cardinal angles are broadly rounded and obtuse. The place of maximum width is commonly slightly posterior to midlength. Small specimens are completely without lobation, but after two or three centimeters of growth a somewhat squared ventral fold becomes developed. On larger specimens it is not well elevated above the anterolateral flanks, but is expressed most strongly by a trapezoidal deflection of the commissure which accommodates a prominent tonguelike extension of a median dorsal sulcus. The latter is broad, shallow, and flat-bottomed. The anterior commissure is not thrown into any additional flexures by the presence of furrows or plications. The exterior is completely smooth except for very rare concentric growth lines on a few specimens and except for faint long plications on the midregions of a few of the largest specimens.

Interior of pedicle valve.—The hinge teeth are very small and delicate, slightly elongate in a direction nearly parallel to the plane of symmetry. The delthyrium opens ventrally into a spondylium of rhomboidal outline and deep V-shaped cross section which is supported by a stout, bladelike median septum that commonly extends past midlength. Its anterodorsal edge is concave and the whole septum may be more or less buried in variable deposits of secondary shell material that can be very thick in the umbonal regions and extend some distance anteriorly. The interior of the valves is smooth.

Interior of brachial valve.—The sockets are shallow, groovelike, and broadly divergent. They join curving, triangular inner plates which in turn join platelike, divergent outer plates which meet the base of the valve along slightly diverging paths as they proceed anteriorly. The interior of the valves is smooth.

Comparison.—The only named species that appears to be close to the new subspecies described here as *G. pelagica lux* is the brachiopod called *Gypidula olga* Khodalevich, 1939. It is the type species of the genus *Levigatella* Andronov, 1961. If it could be proven that *Levigatella* is a valid genus the new subspecies, at least, would certainly be assigned to it, but the validity of *Levigatella* is doubtful. The new subspecies shows a good deal of similarity to *Gypidula olga,* but that species is too poorly illustrated for satisfactory comparison. Consequently such a test must be reserved for the future. The new subspecies includes specimens from the Gedinnian at Cathedral Mountain, Northwest Territories, whose similarity was noted above.

Occurrence.—F fauna; USNM 10794, and UCR 5459, 5460, 5461, 5463, and 5468, all in the Willow Creek area. Also UCR 5473 in the Birch Creek area, and UCR 5457 and 5458 at Birch Creek section II–III.

Figured specimens.—USNM 157061, 171465–171472.

Gypidula and *Sieberella* spp.
(Pl. 19, figs. 13–18)

Discussion.—There are a number of collections from the Pridolian-Gedinnian interval which contain smooth or nearly smooth specimens of *Gypidula.* Generally there are only a small number in each collection and they are poorly preserved. Some or many of these collections may be of *Gypidula pelagica,* but that cannot be decided because the specimens are commonly small and/or have their anterior portions broken away. In addition, because of the lack of a good collection from each locality that yields some specimens, not all those listed below certainly belong to *Gypidula* rather than *Sieberella* (the illustrated specimens illustrate *Sieberella* interiors).

Two localities of those listed below, UCR 5462 and 5448, contain some relatively well-preserved specimens that appear not to belong to *Gypidula pelagica* and have a narrower fold that tends to be bilobate anteriorly.

Material.—A total of 305 silicified specimens from twenty-seven different localities.

Occurrence.—E and F faunas; USNM 10795, and UCR 5461, 5462, 5464, 5465, and 5467, all in the vicinity of Willow Creek. Also UCR 5428, 5430, 5434, 5436, 5437, 5441, 5443, 5444, 5445, 5446, 5447, 5448, 5449, 5450, 5451, 5452, 5453, 5454, 5455, and 5456 at Birch Creek section II–III. Also UCR 5469, at 1,670 ft, and UCR 5470, at approximately 1,870 ft in the Pete Hanson Creek section.

Figured specimens.—USNM 171473–171478.

Gypidula sp. F
(Pl. 20, figs. 1–7)

Gypidula sp. Lenz, 1970:487, pl. 84, figs. 16–22, 24.

Material.—Twenty-seven silicified specimens.

Discussion.—The available specimens somewhat resemble those illustrated as *Sieberella* cf. *problematica* by Johnson (1970) from the slightly younger *Quadrithyris* Zone. The specimens at hand, however, do not have the distinctive biplicate fold of *S.* cf. *problematica.* The available specimens are small and the plications begin well anterior to the umbo and include three on the ventral fold plus a pair of lateral plications adjacent to the fold. Internally, a spondylium and median septum are present in the pedicle valve. In the brachial valve the inner plates diverge slightly anterolaterally while their bases turn sharply toward the midline just above the base of the valve to join along a single track at the midline. The specimens come from F fauna localities UCR 5469 at 1,670 ft and UCR 5470 at approximately 1,870 ft, Pete Hanson Creek and from UCR 5461 and UCR 5466 west of Willow Creek.

Figured specimens.—USNM 171479–171481.

Order STROPHOMENIDA
Suborder STROPHOMENOIDEA

Discussion.—Strophomenid brachiopods are rare at most horizons in the Pridolian-Gedinnian interval. *Leptaenisca* occurs in a few collections in the F fauna of Gedinnian age and several species of *Leptaena* occur in both stages. *Lepidoleptaena* has a rare occurrence in the upper F fauna. An important occurrence of the strophomenid group is represented by the orthotetacean genus *Morinorhynchus* which richly endows the E fauna of Pridolian age. *Iridistrophia,* "*Schuchertella,*" and *Aesopomum* occur in beds of Gedinnian age, but are uncommon. The stropheodontid genera are very poorly represented and include rare *Leptostrophia* and *Strophonella* in the F fauna. The distinctive species *Mesodouvillina costatuloides* is an important and easily recognized member of the E fauna of Pridolian age and a similar form extends rarely into the Gedinnian.

Superfamily Strophomenacea King
Family Strophomenidae King
Subfamily Leptaenoideinae Williams
Genus <i>Leptaenisca</i> Beecher
Leptaenisca sp.
(Pl. 20, figs. 8–11)

Material.—Eleven silicified specimens from three localities.

Exterior.—The valves are subquadrate, shield-shaped in outline, and plano-convex in lateral profile, with a deep pedicle valve which attains its depth through a long geniculate trail. Posteriors of pedicle valves available are too poorly preserved to serve as the basis for a description. Brachial valves have a straight hinge line that equals the maximum width of the valve and a catacline to hypercline, ribbonlike interarea. Pedicle valves have a large, rough flat area formed as a cicatrix of attachment, commonly occupying the major part of the disc from which the geniculate trail proceeds. The latter may be as long as the length of the pedicle valve. No external ornament was observed.

Interior of pedicle valve.—There is a prominent pair of muscle-bounding ridges enclosing a small apical area of muscle impression, of trapezoidal outline. That area is cleft medially by a strong, bladelike ridge which joins anteriorly with a low, rounded ridge that transects the remainder of the shell to the anterior margin in some specimens. The interior is otherwise smooth.

Interior of brachial valve.—Only two relatively well-preserved brachial valves are available and these are both illustrated. They show some variation in the region of the cardinalia which might not be expected. In one specimen the cardinal process lobes are very small and situated posterior to a subtriangular pair of adductor scars, but in the other specimen the cardinal process lobes are quite ponderous, occupying a relatively larger portion of the interior. Both of these specimens show the broadly developed spiraling impressions of brachial ridges proceeding laterally from the anterior edge of the muscle impressions toward the margins of the valve and curving inwardly toward the midline in an ever-decreasing spiral. The interior of both these specimens is strongly papillose, indicating a pseudopunctate shell structure.

Occurrence.—F fauna; UCR 5454 at Birch Creek section II–III, UCR 5462 at Willow Creek, UCR 5469 at 1,670 ft, Pete Hanson Creek.

Figured specimens.—USNM 171482–171484.

Subfamily LEPTAENINAE Hall and Clarke

Genus *Leptaena* Dalman

Leptaena sp. E

(Pl. 1, figs. 22–26)

Diagnosis.—A small, quadrate species of *Leptaena* with concentric and radial ornament of about equal strength; brachial valve with geniculate rim deflected ventrally.

Material.—Ten silicified specimens from two localities.

Exterior.—The valves are small and subquadrate in outline and plano-convex in lateral profile. The hinge line is long and straight, but except in unusual examples the cardinal angles are obtuse so that maximum width is anterior to the hinge line. The ventral interarea is low, triangular, flat, and apsacline. It is cleft medially by a broad, triangular delthyrium which, on the only relatively well-preserved specimen, is only closed apically by deltidial structures. The beak is pierced by a prominent circular foramen. The pedicle valve is geniculate dorsally around the lateral and anterior margin and the anterior margin tends to be straight or flattened and depressed medially at the trail. Brachial valves have a marginal rim geniculate ventrally.

Both valves have an ornament of concentric rugae and radial costae of about equal strength, producing a semireticulate pattern that gives the shells a distinctive appearance.

Interior of pedicle valve.—Hinge teeth are blunt and triangular in cross section with the long edges of the triangle parallel to the hinge line. Dental lamellae are not developed and the diductor impression is subcircular, situated apically, and strongly bordered laterally and anteriorly by elevated marginal ridges. The diductor scar is bisected medially by a rectangular, elevated, longitudinal ridge. The interior is relatively smooth although concentrically corrugated at a few intervals, reflecting the external ornament. In addition, the radial bifurcating impressions of the mantle canal system may be developed.

Interior of brachial valve.—Sockets are broadly set apart anterolaterally diverging, pyriform grooves set adjacent to the posterior margin. Between them is a pair of cardinal process lobes that are nearly conjunct posteriorly and which diverge anterolaterally. Adductor scars and a breviseptum are present, but not easily characterizable because of relatively poor preservation of the available specimens. Interiors appear to be smooth on the available specimens.

Comparison.—*Leptaena* sp. E is distinguished from other Silurian and Lower Devonian species of *Leptaena* from the Great Basin in its small size, deep body cavity, and reticulate ornament.

Occurrence.—E fauna; UCR 5430 and 5429 at Birch Creek section II–III.

Figured specimens.—USNM 171283, 171284.

Leptaena sp. F
(Pl. 20, figs. 12–14)

Material.—Twenty-five silicified specimens from a single locality and a small number of additional specimens tentatively identified with this species.

Exterior.—These are medium-sized *Leptaena* of transverse, shield-shaped outline. The long, straight hinge line is the place of the maximum width with acute, sometimes auriculate, cardinal angles. In lateral profile the valves are plano-convex with only a gently curved pedicle valve attaining depth of body cavity for the most part through anterior geniculation. The visceral disc might be best described as subplanar. The ventral interarea is extremely low, long, and nearly linear rather than triangular. It is flat and apsacline. The dorsal interarea is flat, nearly linear, and anacline. Deltidial structures are not well preserved, although it is clear that the chilidium filled the delthyrium of the pedicle valve.

The exterior is relatively coarsely rugate and radial costae must have been relatively fine because they are not exhibited owing to poor preservation. The concentric rugae number six to eight on most specimens. Both valves have a short trail geniculate dorsally.

Interior of pedicle valve.—Hinge teeth are stubby and triangular with their long edges parallel to the hinge line. Dental lamellae are absent and the muscle scar is defined laterally by strongly elevated muscle-bounding ridges. The latter do not extend around the anterior margin of the muscle scar, but are breached by a pair of longitudinal grooves laterally, adjacent to an angular ridgelike myophragm that bisects the muscle scar.

Interior of brachial valve.—Poor preservation prohibits an accurate description except to note that the cardinal process lobes are typical of the genus, being closely set together posteriorly and diverging anteriorly. Muscle scars are poorly impressed and there is a short, low breviseptum near midlength.

Comparison.—*Leptaena* sp. F is more nearly planar and has coarser concentric rugae and only poorly defined radial ornament compared with other species of *Leptaena* in the Great Basin Silurian and Gedinnian interval. There is considerable resemblance to the Lower Devonian *Spinoplasia* Zone species figured by Johnson (1970) and called *Leptaena* cf. *acuticuspidata,* as well as younger forms in the same series from the *Trematospira, kobehana,* and *Eurekaspirifer pinyonensis* zones. The *Spinoplasia* Zone form appears to be even more coarsely rugate than *Leptaena* sp. F, but specimens from the *kobehana* and *pinyonensis* zones are rather similar. The main distinction at present seems to be that *Leptaena* sp. F is more transverse compared with the younger Early Devonian specimens which are more nearly equal in length and width.

Occurrence.—F fauna; UCR 5458 at Birch Creek section II–III. Specimens regarded here as *Leptaena* cf. sp. F, but which appear to have slightly more numerous concentric rugae, occur at UCR 5462 near Willow Creek, and UCR 5441 and 5443 at Birch Creek section II–III. Poorly preserved specimens assigned to *Leptaena* sp. occur in UCR 5461 and 5466 near Willow Creek and UCR 5442 and 5444 at Birch Creek section II–III. Also UCR 5473 at Birch Creek away from the measured section.

Figured specimens.—USNM 171485, 171486.

Genus *Lepidoleptaena* Havlíček
Lepidoleptaena sp.
(Pl. 20, figs. 15, 16)

Discussion.—There are only ten fragmentary silicified specimens from two localities. These show an auriculate form of the valves with a tendency to elevation of the pedicle valve margins anterolaterally. Internally, pedicle valves have the high ridges that border the disk of the pedicle valve, as in the type species of *Lepidoleptaena,* but these ridges are set more widely from the midline than in the type species (Havlíček, 1967, fig. 45). In addition, brachial valves have the internal diaphragm peculiar to the genus, but it also is more widely set than the type species. The exteriors appear to be almost smooth, but the preservation of the silicified specimens is rather poor. A single fragment of brachial valve does show the long, strong brachiophore ridges that Havlíček has noted in *Lepidoleptaena lepidula.*

Occurrence.—F fauna; UCR 5457 and 5458 at Birch Creek section II–III.

Figured specimens.—USNM 171487, 171488.

Superfamily DAVIDSONIACEA King
Family CHILIDIOPSIDIDAE Boucot

Diagnosis.—Impunctate davidsoniaceans.

Genus *Morinorhynchus* Havlíček, 1965
Type species.—*Morinorhynchus dalmanelliformis* Havlíček, 1965.

Morinorhynchus punctorostra n. sp.
(Pl. 4, figs. 1–20; pl. 5, figs. 1–17)

Diagnosis.—*Morinorhynchus* with ventral beak pierced by throughgoing holes; shell shape variable.

Material.—A total of 894 silicified, disarticulated specimens.

Exterior.—The ventral beak is unusually constructed. It is invariably pierced by a number of small circular holes whose locus is a pair of radial tracks diverging away from the apex of the ventral beak and extending a short distance anteriorly. The shape of the shells is variable although generally width is greater than length. Pedicle valves vary from suboval to subquadrate or subpentagonal and brachial valves vary from subsemicircular to subquadrate. The valves vary in lateral profile from almost plano-convex to strongly biconvex. Nearly flat brachial valves may bear a shallow and indistinct sulcus that causes the anterior commissure to arch ventrally. Some brachial valves are very deeply convex and globose and these lack sulcation. Pedicle valves vary from low, gently convex, "typical" strophic shells with a prominent but still relatively low, flat triangular interarea to subconical, deeply convex shells with a high triangular interarea. Twisting of the ventral palintrope is common, especially on the valves that have a high triangular interarea. Asymmetrical twisting to the left and to the right both occur. The hinge line is long and straight and may be the place of maximum width for some shells with cardinal angles that are approximately right angles. In most cases cardinal angles are slightly obtuse and maximum width is attained near midlength. The ventral interarea is flat or slightly curved and is apsacline, tending toward the catacline position. The delthyrium is completely covered by an outwardly convex pseudodeltidium. The notothyrium lacks a chilidial covering. The dorsal interarea is flat, broad, prominently developed, and anacline.

The ornament consists of numerous radial costellae that increase in number anteriorly by intercalation. In some specimens first-order costellae become slightly enlarged in width relative to later-appearing adjoining costellae, producing a faint parvicostellation.

Interior of pedicle valve.—The apex of the delthyrial cavity is open to the shell exterior via the paired radial rows of shell-piercing holes that traverse the umbo of the valve. These two rows of pores generally lie closely adjacent to the medial edges of the dental plates and do not occur lateral to them. The hinge teeth are stubby, round, inconspicuous projections supported by dental lamellae whose length is approximately equal to the height of the palintrope. The dental lamellae are composed of two parts, a main part adjoining the tracks of the hinge teeth and bowing slightly laterally in section view. These main parts are joined to the floor of the valve by a pair of flat, divergent ridges (pl. 4, fig. 10). Muscle scars are generally inconspicuous, but where preserved the adductor scars are seen to be elongate-oval and situated apically without a prominent division between halves. Well-preserved adductor scars have a very rough surface, attesting to the presence of shell material of a distinctive texture (pl. 4, fig. 20). The diductor scars extend beyond the distal edges of the dental lamellae and cover a broad area, but without flabellation or any markings to accurately outline the anterior edge of the muscle impressions. Margins of the valve tend to be deeply crenulated with the individual ridges fading toward the posterior as they are covered by secondary shell material.

Interior of brachial valve.—The socket plates curve around in a cylindroidal fashion from the posterolateral portions of the dorsal interior and have thickened ventral edges serving as brachiophores. Socket plates are bridged medially by a flat cardinal plate on which there is a pair of posteriorly facing diductor ridges divided by a medial plate. In convex brachial valves there is a deep cavity beneath the cardinalia without a myophragm or any differentiation of adductor muscle scars. Peripheral corrugation may be strong, disappearing posteriorly by the addition of secondary shell covering. The ridges are for the most part rounded or tubular and simple, but these may be medially grooved or may bifurcate.

Comparison.—*Morinorhynchus punctorostra* differs from other named species, *M. dalmanelliformis* Havlíček and *M. attenuata* (Amsden) in having many individuals with a high semiconical pedicle valve and a convex brachial valve. The three first mentioned species are all relatively simple lenticular shells with low triangular interareas. In addition, *Morinorhynchus punctorostra* is apparently unique in the condition of its perforated ventral beak.

Discussion.—The perforated ventral beak of *Morinorhynchus punctorostra* evidently is an artifact of the mode of attachment of the shell. The common occurrence of pedicle valves with a high, triangular, twisted palintrope attests to their having been attached, but the complete imperforate pseudodeltidium blocked any pedicle exit via the posterior of the valve. Furthermore there is no cicatrix of attachment, nor is there a single circular perforation as is typical of *Leptaena*. It appears that *Morinorhynchus punctorostra* attached through a number of fleshy threads that were isolated and restricted to a number of separate perforations via growth along the anterior margin at an early growth stage.

Occurrence.—E fauna; the illustrated types and a large collection of paratypes are from UCR 5430 at Birch Creek section II–III. Other occurrences are in UCR 5429 and 5432 at Birch Creek section II–III. In addition, there are seventeen specimens in UCR 5433 and fifteen specimens in UCR 5431 at Birch Creek section II–III.

Figured specimens.—USNM 171314–171333.

Genus *Iridistrophia* Havlíček, 1965

Type species.—Orthis umbella Barrande, 1848.

Iridistrophia cf. *umbella* (Barrande)

(Pl. 21, figs. 1–8)

Orthis umbella Barrande, 1848:206, pl. 19, fig. 1.
Orthis umbella Barrande, 1879, pl. 58, fig. 1.
Strophomena subtilis Barrande, 1879, pl. 51, case 1, according to Havlíček, 1967:194.
Schellwienella praeumbracula Kozlowski, 1929:105, text fig. 32; pl. 5, figs. 3–6.
Schellwienella praeumbracula Nikiforova, 1954:84, pl. 8, fig. 5.
Iridistrophia umbella Havlíček, 1965:292, pl. 1, figs. 4–6, 9, 11.
Iridistrophia umbella Havlíček, 1967:194, pl. 41, figs. 7–16; pl. 42, figs. 3, 5, 8.

Discussion.—The specimens from Nevada, although fragmentary, appear to be relatively close to the Bohemian and Podolian forms. The principal difference appears to lie in the stronger curvature of the costae near the posterolateral extremities of the Nevada specimens. In the Bohemian specimens illustrated by Havlíček (1967, pl. 41) the costae nearest the hinge line almost parallel it and are nearly straight. The hinge line is also consistently longer and is commonly, but not always, the place of maximum width. The Podolian specimens, called *praeumbracula* by Kozlowski are not well illustrated, but on the basis of the available figures they appear to be indistinguishable from the Bohemian specimens. Posterolateral costae are nearly straight in both and a better suite of shells from Podolia might reveal more specimens with long hinge lines and cardinal angles that are near 90°.

Material.—Twenty-nine silicified specimens from a single locality.

Exterior.—The valves are rounded-subquadrate in outline and resupinate in lateral profile. The hinge line is straight and equal to about three-fifths to two-thirds the maximum width. The ventral interarea is flat, low, and triangular and is cleft by a large triangular delthyrium. The specimens are poorly preserved and in most there is no deltidial covering, but in one poorly preserved specimen there is an outwardly convex pseudodeltidium filling the apical half of the delthyrium. The dorsal interarea is about the same length as the ventral interarea and is ribbonlike and catacline. The chilidium is only a small ridge crossing the base of the cardinal process lobes posteriorly. Cardinal angles are well rounded and obtuse and maximum width is between the hinge line and midlength.

The ornament consists of numerous, subangular to rounded, radial costellae that increase in number anteriorly by intercalation. There is no indication of parvicostellation on most specimens, but one brachial valve (pl. 21, fig. 5) shows a slight tendency in this direction. The radial disposition of the costellae breaks down lateral to the midregions of the valve. On the posterolateral portions of the flanks the costellae become progressively more and more curved, concave toward the hinge line, so that a few costellae intersect the beak ridges defining the ventral edge of the palintrope.

Interior of pedicle valve.—Muscle scars are not impressed and the hinge teeth are too poorly preserved to allow accurate description. They are, however, supported by short dental plates that extend anterolaterally almost to the base of the teeth. The interior is smooth to faintly ridged and radially grooved over most of the interior. Radial grooving is strong peripherally.

Interior of brachial valve.—The socket plates are very broadly divergent, defining shallow, pyriform sockets between them and the posterior margin of the valve. The posteromedial portions of the socket plates are strongly twisted or curved posteriorly where they join a pair of cardinal process lobes that project ventrally and face posteriorly. The paired myophores are each cleft nearly parallel to the plane of symmetry, forming a quadrilobate myophore. There is no myophragm at the base of the valve. The interior is radially ridged and marginally crenulated as in pedicle valves.

Occurrence.—F fauna; UCR 5458 at Birch Creek section II–III.

Figured specimens.—USNM 171492–171494.

Genus *"Schuchertella"*

Discussion.—As noted elsewhere and discussed by Johnson (1970) the type species of *Schuchertella* Girty is pseudopunctate. Many otherwise similar Lower and Middle Devonian schuchertelloid species therefore lack a generic name. Specimens that fall in this group are listed here as *"Schuchertella"* sp.

"Schuchertella" sp.

Discussion.—There are only eighteen silicified specimens representing *"Schuchertella"* in collections from the Roberts Mountains Formation. These are inadequate on which to base a full description. The specimens are small or medium-sized, ranging up to a maximum dimension of about an inch. They are costellate and are lenticular, biconvex shells without a high palintrope in any of the specimens. Several specimens showing the ventral interior reveal very small, widely divergent, strutlike dental lamellae. Of interest is the fact that the pedicle valves have an inverted V-shaped hole or slot at the apex at the point of attachment. This is similar to what occurs in *Morinorhynchus punctorostra*, but in that species, more commonly, there is a series of small holes which together form an inverted V-shaped arrangement and only in a few specimens are they so closely spaced as to merge into a single V-shaped slot.

Occurrence.—F fauna; UCR 5461 and 5462, both at Willow Creek; also UCR 5466, west of Willow Creek. Also five specimens from UCR 5473 in the Birch Creek area.

Genus *Aesopomum* Havlíček, 1965

Type species.—*Strophomena aesopea* Barrande, 1879.

Aesopomum varistriatus Johnson, 1970
(Pl. 20, figs. 17–23)

Aesopomum varistriatus Johnson, 1970:111, pl. 17, figs. 9–14.

Discussion.—This fossil is invariably rare and preserved only as fragmentary

specimens in several collections at hand. Specimens of *Aesopomum* were illustrated originally by Johnson (1970) from several different localities and horizons
and the name *Aesopomum varistriatus* was given to specimens from the Gedinnian of the Roberts Mountains Formation. Younger specimens, for example, from
the *Quadrithyris* Zone of the Windmill Limestone, appeared to differ from *A.
varistriatus* in having the paired cardinal process lobes more or less as continuous
extensions of the socket ridges. This distinguished them from *A. varistriatus*
which has the socket plates and the cardinal process lobes (or lobe if fused
medially) discrete and separate, although connected posteriorly.

   Material.—Twenty-five silicified specimens.

   Exterior.—The shells have a variable outline and, considering the generally
poorly preserved condition of almost all of them, no good characterization of the
outline is possible. In lateral profile they are deeply biconvex with a well-rounded
brachial valve and a conical or subpyramidal pedicle valve, commonly with a
high flat triangular interarea that may be irregular and asymmetrical. The delthyrium is covered by an elongate, outwardly convex pseudodeltidium. The
exteriors are concentrically rugose and finely radially costellate.

   Interior of pedicle valve.—The teeth are elliptical in cross section with their
long axes tending parallel to the plane of symmetry. They are not supported by
dental lamellate, but instead proceed as ridgelike tracks to the apex of the valve.
No muscle scars are visible and the interior of the shell is generally smooth,
without any irregular papillose development that might indicate the presence of
pseudopunctae.

   Interior of brachial valve.—The cardinalia consist of socket plates and a platelike cardinal process that extends ventrally as a single plate that was formed from
a pair of fused plates, the bilobed nature generally being exhibited by the medially grooved distal end. The platelike cardinal process joins at its proximal-lateral
parts with the posteromedial edges of the socket plates which project anterolaterally into the shell. Muscle scars were not observed.

   Occurrence.—F fauna; UCR 5462, west of Willow Creek. Also UCR 5443 and
UCR 5445, both at Birch Creek section II–III. *Aesopomum* sp., possibly *A. vari-
striatus* but poorly preserved, occurs in UCR 5465 west of Willow Creek and in
UCR 5442, 5444, and 5453, at Birch Creek section II–III. Questionably identified
Aesopomum occurs in UCR 5431 at Birch Creek section II–III.

   Figured specimens.—USNM 171489–171491.

Superfamily STROPHEODONTIDAE Caster
Family LEPTOSTROPHIIDAE Caster
Genus *Leptostrophia* Hall and Clarke
Leptostrophia? sp.

   Discussion.—There is a single small silicified specimen with articulated valves.
This specimen is subplanate and nearly semicircular, with very fine radial costellae. The interiors are not visible. The specimen may belong to *Leptostrophia,* but
this cannot be determined with the available material. The single specimen comes
from UCR 5460, Willow Creek.

   Specimens confidently assigned to *Leptostrophia* occur in the Pete Hanson
Creek F fauna collection UCR 4449, a collection not fully treated in this report.

Family STROPHEODONTIDAE Caster
Genus STROPHONELLA Hall
Strophonella sp.

Material.—Thirty-two fragmentary silicified specimens.

Exterior.—The valves are of medium size and subsemicircular in outline with resupinate lateral profile. Cardinal angles are pointed and acute and the hinge line is the place of maximum width. The ventral interarea is flat and apsacline.

The ornament consists of a broadly parvicostellate arrangement in which fine radial costae are not observable between the primaries. Whether this is the true condition of the exteriors, however, is uncertain because of poor preservation.

Interior of pedicle valve.—Only a single specimen shows the interior of the pedicle valve. It has a broadly flabellate diductor scar that extends almost half-way to the anterior margin. Its posterolateral margins are almost straight rather than being markedly concave toward the posterolateral extremities.

Interior of brachial valve.—The cardinalia consist of a pair of discrete, plate-like socket plates that become somewhat swollen and rounded on their distal ends. They are directed ventrally and slightly posteriorly and appear to have formed from subparallel platelike bases. The brachiophore plates are set widely lateral to the cardinal process lobes and diverge anterolaterally at a broad angle. Neither myophragm nor muscle scars are developed on the available fragmentary specimens. Small specimens show a radial corrugation reflecting the external ornament, but the larger specimens are smooth internally, although preservation is too crude to allow for the determination of small pustules that generally indicate the presence of pseudopunctae.

Occurrence.—F fauna; UCR 5444 and 5458 at Birch Creek section II–III. Also UCR 5473 in the Birch Creek area. *Strophonella* sp. also occurs in UCR 5461 and 5462, west of Willow Creek.

Family DOUVILLINIDAE Caster
Genus *Mesodouvillina* Williams
Mesodouvillina costatuloides n. sp.
(Pl. 6. figs. 1–18)

Diagnosis.—Small, transverse, strongly corrugated *Mesodouvillina.*

Material.—Forty-three topotype specimens and fourteen additional specimens, all silicified.

Exterior.—The valves are small and transverse with an outline that is nearly semicircular. The valves are concavo-convex in lateral profile with only moderate pedicle valve curvature. The hinge line is long and straight and is the place of maximum width. Cardinal angles commonly are acute and auriculate. The ventral interarea is long, flat, and bandlike with an inclination that is apsacline, approaching the orthocline position. The dorsal interarea is ridgelike and linear. The interarea is denticulate along slightly more than half its length.

The ornament consists of a parvicostellate arrangement with strongly raised primaries separating more numerous very fine costellae, together with well-developed concentric rugae that are discontinuous between the primary radial costellae. On anterior parts of the shell the rugation is seen to be associated with a netwook of shell-piercing pores that are situated in the furrows between ridges of the rugate pattern (pl. 6, fig. 16). The posterolateral extremities of the shells are devoid of radial ornament.

Interior of pedicle valve.—The muscle field is roughly triangular and is defined by strong, almost platelike, muscle-bounding ridges that diverge at an angle slightly less than 90°. On some specimens these are deflected medially near their midlength to enclose a rhomboidal muscle field, but in no specimen is the anterior edge of the muscle field elevated on a platform. The diductor field is bilobed but not flabellate, and each half of the diductor field may be divided into inner and outer portions. The adductor scars are situated well in the posterior half of the muscle field and consist of a pair of small, elongate lobes set closely together at the midline of the valve. The shell material was generally thin so that the rugate pattern and the primary costellae are reflected on the internal surface of the valves.

Interior of brachial valve.—Brachiophore plates are situated well to the posterior and are widely divergent away from cardinal process lobes which are situated medially and which project posteroventrally. The bases of the cardinal process lobes converge and join on a broad, low, rounded ridge between the adductor scars. The adductor scars are bounded posterolaterally by a pair of broad, low, rounded ridges. There is a pair of slightly divergent, low, rounded brace-plates projecting anteriorly slightly short of the midlength of the valve from the area between the inner margin of the adductor scars and the lateral edges of the rounded ridge that divides them. The interior is lightly crenulated, reflecting the external rugation and primary costation.

Comparison.—*Mesodouvillina costatuloides* resembles only one other named species, that is *Mesodouvillina costatula* (Barrande). They are linked by the presence of a corrugated shell and the peculiar rhomboidal ventral muscle field (cf. Havlíček, 1967, pl. 34, fig. 11). The new species differs principally in being transverse rather than equidimensional, or even elongate.

Occurrence.—E fauna; the types of *Mesodouvillina costatuloides* are from UCR 5430 at Birch Creek section II–III. Hypotypes are from UCR 5429 and 5431, Birch Creek section II–III. Rare specimens of *Mesodouvillina* cf. *costatuloides* occur in the F fauna at UCR 5459 and 5462, west of Willow Creek; also UCR 5442 and 5445 at Birch Creek section II–III.

Figured specimens.—USNM 171334–171342.

Order RHYNCHONELLIDA
Suborder RHYNCHONELLOIDEA

Discussion.—Johnson (1970:141) noted that the rhynchonellids of the Lower Devonian of Nevada were marked by little phyletic continuity and seem to appear and disappear from the stratal sequence without obvious precursors or descendants. The same is true of the Pridolian-Gedinnian interval of the Roberts Mountains Formation. The genus *Ancillotoechia* is abundant in a few collections of Gedinnian age, but is absent from most. *Sphaerirhynchia* also occurs in the Gedinnian, where it is less abundant, but it occurs in many collections. So far *Sphaerirhynchia* has not been found at any other horizon in the Silurian or Devonian of Nevada. The next higher occurrence of *Ancillotoechia* is in the *Trematospira* Zone of the Siegenian. Other rhynchonellid genera that occur in the Pridolian-Gedinnian interval, such as *Hebetoechia, Lanceomyonia,* and a questionable occurrence of *Eoglossinotoechia,* are represented only by a few specimens.

Superfamily CAMAROTOECHIACEA Havlíček
Family TRIGONIRHYNCHIIDAE Schmidt
Genus *Ancillotoechia* Havlíček
Ancillotoechia gutta n. sp.*
(Pl. 22, figs. 1–19)

Diagnosis.—Small shell of subequal length and width; ventral flanks curved, not reflexed; dorsal costae four, dorsal median septum present.

Material.—A total of 132 silicified specimens from two localities.

Exterior.—The shells are subtriangular to pyriform in outline, but with most specimens retaining approximately equal width and length. The apical angle is about 90°, slightly more or slightly less, with posterolateral margins tending to be slightly concave. In lateral profile the valves are subequally biconvex to unequally biconvex with the brachial valve having as much as twice the depth of the pedicle valve. The ventral beak is suberect and protrudes beyond the brachial valve markedly in small specimens, but is closely appressed to the dorsal umbo in larger specimens. The delthyrium is triangular and open. Deltidial plates were not observed. Maximum width of the valves is generally slightly anterior to midlength which is slightly beyond where the posterolateral flanks begin curving strongly toward the anterior midline.

The shells bear simple radial costae, three in number in the ventral sulcus and four in number on the dorsal fold. The fold and sulcus are well differentiated from the flanks, without the intervention of parietal costae, but the ventral sulcus is part of the even curvature of the valve. It is not accentuated by the development of an exaggerated long tongue, nor by the development of reflexed ventral flanks—forming a relatively conservative anterior commissure with a zigzag crenulate margin of interlocking costae. Flanks commonly have five, well-defined, subangular costae divided by V-shaped subangular intercostal furrows. Growth lines or other concentric ornament were not observed.

Interior of pedicle valve.—Hinge teeth are small and delicate and are supported by thin, platelike, slightly diverging dental lamellae. Muscle scars are not impressed. The interior is corrugated, reflecting the costation of the shell.

Interior of brachial valve.—The cardinalia consist of socket ridges and crural bases along the inner edges of the socket ridges, forming a structure lying ventral to a low, V-shaped septalium and septum. The septum extends a little less than to the midlength of the valves. Muscle scars are not impressed and the interior is corrugated, reflecting the costation of the shell. No inner hinge plates were observed covering the septalium.

Comparison.—*Ancillotoechia gutta* bears a relatively close resemblance to *A. nucula* as illustrated from the Skala of Podolia by Kozlowski (1929, pl. 6), but the new species does not contain many of the broad varieties as illustrated by Kozlowski nor are the costae as numerous as on Kozlowski's specimens. There is

---

* This species may prove to be more accurately placed in the genus *Hemitoechia* Nikiforova, 1970.

a considerable external similarity between *Ancillotoechia gutta* and *A. aptata* Johnson, 1970. *Ancillotoechia gutta* has a less extravagantly developed sulcation of the anterior commissure and is, on the average, less lenticular, but has a better-developed median septum in the brachial valve. McLaren, Norris, and McGregor (1962, pl. 8, figs. 7–9) have illustrated an unnamed *Ancillotoechia* species that is similar to *A. gutta* and which may be conspecific.

Occurence.—F fauna; all the specimens come from localities UCR 5459 and 5460 west of Willow Creek. Questionably identified *Ancillotoechia* occur in UCR 5470, Pete Hanson Creek, and in UCR 5466, west of Willow Creek.

Figured specimens.—USNM 171495–171501.

Family UNCINULIDAE Rzhonsnitskaya
Genus *Sphaerirhynchia* Muir Wood and Cooper
Sphaerirhynchia gibbosa (Nikiforova)
(Pl. 23, figs. 1–11)

Wilsonella wilsoni var. *gibbosa* Nikiforova, 1954:110, pl. 11, figs. 5a–d.

Discussion.—The specimens from Nevada agree closely with Nikiforova's illustrated specimen in size, outline, costation, development of the anterior commissure, and in lateral profile except that the depth of articulated valves is relatively greater in the Podolian specimen. This should be a feature of some intraspecific variation, so there seems to be no way to separate the specimens from Nevada from those from Podolia, at least until other Podolian specimens are illustrated for comparison.

Material.—Thirty-four silicified specimens from ten localities.

Exterior.—The valves are pyriform in outline and strongly subequally biconvex in lateral profile. The ventral beak is strongly incurved and closely appressed to the dorsal umbo. The apical angle is about 90° with posterolateral margins nearly straight to about midlength, which is the point of maximum width. The anterior outline of the shell is subsemicircular. In transverse sections the articulated valves are suboval, modified slightly to a quadrate configuration, and are wider than deep. The form of articulated valves is subcuboidal, modified in the anterior half by a broad, shallow, ventral sulcus and by a low, poorly developed, flat-topped dorsal fold. The anterior commissure is moderately to strongly deflected forming a ventral tongue.

The exterior bears relatively numerous simple, low, rounded costae separated by shallow, narrow, V-shaped interspaces. There are five or six costae in the ventral sulcus and as many as ten on each flank. Anteriorly, the costae flatten out and are medially grooved so that interlocking of the two valves is accomplished by protruding points from the intercostal furrows and V-shaped indentation in the costae themselves.

Interior of pedicle valve.—Hinge teeth are small and knoblike and are supported basally by thin, convergent, platelike dental lamellae that change their slope and tend to diverge slightly toward the base of the pedicle valve. Their ventral edges are prolonged as diverging ridges that enclose a narrow, elongate, undifferentiated ventral muscle field. The interior is crenulated, reflecting the external costation.

Interior of brachial valve.—Sockets are delicate cylindroidal grooves set along the posterolateral margins of the valve. A pair of triangular outer hinge plates extend roughly parallel to the commissure and join the crural bases which are supported by a short, V-shaped septalium and a bladelike median septum. Inner hinge plates are present in the anterior portion of the cardinalia. The geometry of the crura was not observed although the crura appear to be bandlike, exceedingly delicate, and unflanged. The interior is slightly crenulated, reflecting the external costation.

Occurrence.—F fauna; *Sphaerirhynchia gibbosa* occurs at localities UCR 5461 and 5462, west of Willow Creek, and at UCR 5458 at Birch Creek section II–III. Also UCR 5473 in the Birch Creek area. *Sphaerirhynchia* sp. indet. occurs at locality UCR 5459, west of Willow Creek; also UCR 5454, 5455 and 5457 at Birch Creek section II–III. *Sphaerirhynchia?* sp. indet. occurs in USNM 10795, east of Willow Creek and 5447, Birch Creek section II–III.

Figured specimens.—USNM 171502–171506.

Genus HEBETOECHIA Havlíček
Hebetoechia? cf. *ornatrix* Havlíček
(Pl. 23, figs. 12–16)

Rhynchonella princeps Barrande, 1879 (in part), pl. 120, case 8, fig. 2.
Hebetoechia ornatrix Havlíček, 1961:121, text fig. 45, pl. 8, figs. 2, 3.

Discussion.—There is only a single pedicle valve available. It is very gently curved posteriorly and becomes slightly reflexed anterolaterally, so that the whole shell is nearly flat. There is a very shallow, broad, sulcus beginning at about mid-length. The sulcus extends as a geniculate tongue anteriorly. Costae are very low and rounded with narrow, V-shaped interspaces. There are ten costae on the sulcus anteriorly and about as many on each flank of the valve. Near the end of the sulcal tongue the costae are flattened and medially grooved.

The interior shows no muscle markings. There are only some radial grooves anteriorly, reflecting the costation of the shell.

Occurrence.—F fauna; UCR 5458 at Birch Creek section II–III.
Figured specimen.—USNM 171507.

Genus *Lanceomyonia* Havlíček
Lanceomyonia cf. *confinis* (Barrande)

Rhynchonella confinis Barrande, 1879, pl. 139, case 1.
Rhynchonella barbara Barrande, 1879, pl. 117, case 7, fig. 1.
Rhynchonella altera Barrande, 1879 (in part), pl. 26, figs. 1, 2.
Lanceomyonia confinis Havlíček, 1961:116, text fig. 43, pl. 13, figs. 4, 5.

Discussion.—There are only nine poorly preserved and fragmentary silicified specimens, from a single locality, representing this taxon. They are insufficient on which to base a full description. The specimens are dorsi-biconvex and are coarse-ribbed and subcuboidal with the costae prominent only on the anterior portions. Internally there are very short dental lamellae, but no discernible muscle scars present. The cardinalia consist of small, discrete, triangular hinge plates without a connecting plate. There is a thin bladelike median septum extending about two-fifths of the distance to the anterior margin. The specimens come from E fauna locality UCR 5430 at Birch Creek section II–III.

Genus *Eoglossinotoechia* Havlíček

Eoglossinotoechia? cf. *C. cacuminata* Havlíček

(Pl. 24, figs. 1–13)

Rhynchonella princeps Barrande, 1879 (in part), pl. 120, cases 2, 3, 4, 6, 11.
Eoglossinotoechia cacuminata Havlíček, 1959b:81.
Eoglossinotoechia cacuminata Havlíček, 1961:162, text figs. 69–73; pl. 26, figs. 1–4.

Material.—Eleven poorly preserved silicified specimens.

Exterior.—The valves are small, pyriform in outline, and unequally biconvex in lateral profile. The pedicle valve is only gently curved but the brachial valve is much deeper. The apical angle was a little more than 90°, with posterolateral flanks extending in about a straight line, then curving around the anterior margin. Maximum width is anterior to midlength. The valves are covered with numerous, small rounded simple costae, divided by relatively narrow interspaces. The costae are flattened and medially grooved for the anterior margin.

Interior.—Ventral structures are not preserved. In the brachial valve the inner hinge plates are triangular, gently concave, disjunct plates with bladelike dorso-medial ridges forming the crural bases. The structure is without septal support, and it is on this point that the generic assignment is questioned. Inasmuch as the few available specimens are very small, however, a satisfactory comparison of structure with the type species is not possible.

Occurrence.—F fauna; UCR 5446 at Birch Creek section II–III; also UCR 5473 in the Birch Creek area and UCR 5461 and 5462 west of Willow Creek. An indeterminate uncinuloid, somewhat resembling this species, occurs in UCR 5430, Birch Creek section II–III. Also from UCR 6306 at Pete Hanson Creek, a collection not otherwise studied for this report.

Figured specimens.—USNM 171508–171513.

Order SPIRIFERIDA
Suborder ATRYPOIDEA

Discussion.—The suborder Atrypoidea has a very important representation in the Silurian in Nevada, exemplified here by the Pridolian. Typical representatives of the genus *Atrypa* occur rarely in Pridolian age beds and commonly in beds of Gedinnian age. The absence of any common true *Atrypa* from the Pridolian age beds is possibly compensated there by the occurrence of *Reticulatrypa*. The genus *Spirigerina*, represented by *S. marginaliformis*, has a single occurrence in the Pridolian-Gedinnian interval in a collection from the basal Gedinnian fauna; possibly its occurrence is limited by ecological considerations and should be regarded as a more typical member of the F fauna. It occurs with F fauna species again in the lower Windmill Limestone at Coal Canyon. The smooth atrypid genus *Dubaria* is represented in Pridolian-Gedinnian faunas by two forms. The Silurian form, represented by *D. megaeroides*, has what might be called an ordinary atrypid cardinalia with simple, thick, curving socket plates. The specimens of *Dubaria* encountered in the Gedinnian fauna have a delicate cardinalia in which the sockets are separately defined from subhorizontal platelike outer hinge plates that lie medial to the sockets. When better material allows demonstration of these differences in the form of adequate illustrations a revised nomenclature must be proposed. Another smooth atrypid genus is *Cryptatrypa*, represented commonly in the Gedinnian fauna by the elongate bullet-shaped *C. angusta. Cryptatrypa* has not been found in the Pridolian fauna; possibly its place there is taken by *Atrypella* which occurs in most Ludlovian and Pridolian age collections, but which is unknown anywhere in beds as young as Gedinnian. The genus *Atrypina* is represented in a number of collections in the Gedinnian, but it is common in none. The genus *Gracianella* occurs in great abundance in Silurian collections from the Roberts Mountains. It has a good representation right up to the Siluro-Devonian boundary, but no *Gracianella* is known from Gedinnian age faunas. Its occurrence in collections without diagnostic brachiopods, or other fossils, frequently is the deciding factor for a Silurian assignment. *Gracianella* is a genus of great variability from species to species and includes both smooth and costate forms. A particularly remarkable plicate species of *Gracianella, G. reflexa* is notable in the E fauna. The genus *Lissatrypa* seems to have a common occurrence in the basal Gedinnian fauna, but it has not been recognized at any other horizon within the Roberts Mountains Formation.

Superfamily ATRYPACEA Gill

Family ATRYPIDAE Gill

Subfamily ATRYPINAE Gill

Genus *Atrypa* Dalman

Atrypa nieczlawiensis Kozlowski

(Pl. 24, figs. 14–27)

Atrypa reticularis var. *nieczlawiensis* Kozlowski, 1929:170, pl. 8, figs. 14–17.
?*Atrypa recticularis* var. *nieczlawiensis* Nikiforova, 1954:118, pl. 12, fig. 6.
Atrypa sp. McLaren, Norris, and McGregor, 1962, pl. 8, figs. 4–6.

Discussion.—Species of the *Atrypa nieczlawiensis* type form a distinct subgroup of *Atrypa* related to the true *Atrypa reticularis.* The former, however, are characterized by narrow, rodlike costellae separated by relatively deep, almost slitlike interspaces and crossed by numerous, regularly spaced growth lines of a fragile nature. All these features of ornament combine with the evenly suboval and biconvex shape of *A. nieczlawiensis* to make it a relatively easily recognized shell. The specimens illustrated by Nikiforova (1954) and ascribed to *Atrypa nieczlawiensis* are queried here because her specimens have a nearly flat pedicle valve and a deeply convex brachial valve. This contrasts to some extent with the specimens originally illustrated by Kozlowski (1929) which are subequally biconvex. Nikiforova's specimens were said to originate from the Malinovetski beds (of Ludlovian age) in Podolia; this is an older horizon than is typical for the species which ordinarily is reported from the Borszczow beds (of Gedinnian age).

In North America a similar species is *Atrypa arctostriata* Foerste as illustrated by Amsden (1949, pl. 9). It seems to be smaller and somewhat more lenticular than *A. nieczlawiensis,* but otherwise is very similar.

Material.—A total of 653 silicified specimens from five localities.

Exterior.—The outline is variable from subcircular to slightly elongate suboval and the valves are subequally biconvex in lateral profile. The ventral beak is very short and incurved, lying closely appressed to the dorsal umbo with only a small circular foramen evident. The posterolateral margins slope evenly around to the place of maximum width, a little posterior of midlength, and this is the place of greatest curvature. The anterior half of the outline is nearly circular. Most specimens lack much development of a fold or sulcus (pl. 24, fig. 14), but a few larger specimens develop a very gentle dorsal deflection of the anterior commissure (pl. 24, fig. 19).

The exterior is covered with numerous rodlike or wirelike, radiating costellae, separated by deep, narrow, almost slitlike interspaces. The radial ornament is interrupted by numerous concentric growth lamellae, each of which appears to have once extended as a separate frill.

Interior of pedicle valve.—Small, oval hinge teeth are present with their axes lying subparallel to the hinge line and separated posteriorly from the valve margins by a pair of grooves. They are supported basally by extremely fine dental lamellae forming small conical umbonal chambers, or these may be completely filled up so that the teeth appear sessile. One well-preserved, thin-shelled specimen examined shows a double set of plates below each tooth in the pedicle valve. That is, the tooth is supported by a curving plate that joins the valve wall and the plate is in turn supported by an additional plate forming a pair of small conical cavities on each side of the valve. Most specimens show no impression of the ventral muscle field, but a few are thick-shelled and show a broad, elongate, impressed scar, joined laterally by narrow pustulose areas.

Interior of brachial valve.—The sockets are broadly divergent with their outer edges defined by the posterolateral valve margins. Medially, they are defined by discrete socket plates onto which are attached short, divergent, pronglike crura that extend into the delthyrial cavity of the pedicle valve. Adductor muscle scars are not impressed and the diductor attachment site is small. A myophragm is commonly lacking. The interior is smooth except for fine radial corrugations, reflecting the costation of the valve.

Occurrence.—F fauna; UCR 5459, 5460, and 5462 at Willow Creek. Also UCR 5457 and 5458 at Birch Creek section II–III. Also UCR 5473 in the Birch Creek area.

Figured specimens.—USNM 171514–171520.

Record of occurrence of *Atrypa "reticularis"*

Material.—A total of 743 silicified specimens.

Occurrence.—E and F faunas; USNM 10795 east of Willow Creek. Also UCR 5461, 5464, 5465, 5466, 5467 west of Willow Creek. Also UCR 5434, 5441, 5443, 5447, 5448, 5449, 5450, 5451, 5452, 5453 5454, 5455, 5456 at Birch Creek section II–III. Also UCR 5469, at 1,670 ft, and 5470 at approximately 1,870 ft, Pete Hanson Creek.

All the specimens listed above are of the *Atrypa "reticularis"* type and many of these may belong to *Atrypa nieczlawiensis;* preservation makes it impossible to determine the assignment accurately. Other specimens clearly do not belong to *Atrypa nieczlawiensis,* but are of the type close to true *Atrypa reticularis.*

Genus *Reticulatrypa* Savage, 1970

Type species.—*Reticulatrypa fairhillensis* Savage, 1970.

Reticulatrypa neutra n. sp.
(Pl. 2, figs. 1–18)

Diagnosis.—Coarsely ribbed *Reticulatrypa* without reversal of ventral fold at anterior commissure.

Material.—Eighty-two silicified specimens.

Exterior.—The shells are transversely suboval in outline and the valves are subequally biconvex in lateral profile. Commonly, the pedicle valve is slightly the deeper of the two. The hinge line is moderately long and only slightly curved, giving some specimens a strophic appearance. Interareas are lacking, but the ventral beak is straight with a broad, low delthyrium closed by deltidial plates except for an apical foramen. Cardinal angles are rounded and obtuse and maximum width is generally posterior to midlength. From the place of maximum width the lateral and anterior margins curve around in an even semicircular fashion. Pedicle valves are subcarinate and have flanks sloping off without curvature from the umbo, but anteriorly the flanks may be slightly convex. Brachial valves may or may not bear a prominent median sulcus that combines with the ventral carination to give many specimens a characteristic transverse profile reminiscent of the genus *Carinatina* Nalivkin.

The exterior is covered with rounded radial costellae or fine costae that are variable in their size and number. The costae increase in number by splitting and by intercalation and are divided by prominent U-shaped interspaces. Bundling occurs on some specimens (pl. 2, figs. 13, 14). The radial ornament is crossed by numerous, lamellose, concentric growth lines which occur at regular intervals over the whole of the valve, but which appear to have been nonfrilly.

Interior of pedicle valve.—The hinge teeth are supported by extremely small, nearly obsolescent dental lamellae. The ventral muscle field is poorly outlined but is elevated anteriorly on a slightly raised platform that drops off precipitiously at its anterior edge into the median furrow that corresponds with the carinate shell configuration. The interior is smooth except for a gentle radial corrugation corresponding with the costation of the valves.

Interior of brachial valve.—The cardinalia consist of simple, discrete, widely divergent socket plates outlining narrow, undivided sockets. A simple or split myophragm may occur in the area between the site of adductor muscle attachment, but the latter is not impressed. The remainder of the interior is smooth except for radial corrugations on thinner-shelled specimens.

Comparison.—Both the type species *Reticulatrypa fairhillensis* and the species *R. losvensis* (Khodalevich, 1951, pl. 22) are more finely ribbed species with an anterior commissure reversal of the ventral carination.

Occurrence.—Pridolian fauna; UCR 5428 and 5437 at Birch Creek section II–III.

Figured specimens.—USNM 171285–171294.

Reticulatrypa aff. granulifera (Barrande)
(Pl. 2, figs. 19–24)

Terebratula granulifera Barrande, 1847:456, pl. 19, fig. 3.
Atrypa granulifera Barrande, 1879, pl. 19, figs. 1a–f; pl. 129, case 5, figs. 1–13.

Material.—Nine silicified specimens.

Discussion.—The material is too poor to describe, but consists of a few small, strongly carinate specimens with numerous fine costellae.

Occurrence.—Pridolian fauna; UCR 5434 and 5436 at Birch Creek section II–III.

Figured specimens.—USNM 171295–171297.

Genus *Spirigerina* d'Orbigny

Spirigerina marginaliformis Alekseeva

(Pl. 25, figs. 1–7)

Atrypa marginalis Khodalevich, 1939:47 (see synonymy), pl. 25, figs. 1–3; not Dalman.
Spirigerina marginaliformis Alekseeva, 1960:65, pl. 7, fig. 1.
Plectatrypa marginalis sibirica Rzhonsnitskaya, 1960, pl. 53, fig. 24.
Spirigerina marginaliformis Alekseeva, 1962:161, text figs. 76, 77; pl. 9, fig. 10.

Discussion.—*Spirigerina marginaliformis* and its synonym *Spirigina sibirica* are more finely ribbed than the species *Spirigerina supramarginalis* (Khalfin). Thus the all-encompassing synonymy of Kulkov (1963) should not be accepted. There is possibility of confusion in Nevada because Johnson (1965:371) originally assigned the *Quadrithyris* Zone specimens to "*Plectatrypa* cf. *P. sibirica.*" More recently (Johnson, 1970) these have been placed in the coarser-ribbed species *S. supramarginalis*.

Material.—Eighty-five silicified specimens.

Exterior.—The valves are commonly slightly transversely suboval, varying to subpentagonal in outline and are unequally biconvex in lateral profile. Brachial valves are more deeply arched than pedicle valves although the latter are also convex. The ventral beak is short, straight, pointed, and pierced apically by a circular subhypothyrid foramen. The hinge line is short and curved; maximum width is generally near midlength. Pedicle valves bear a shallow, broad sulcus of irregular or variable development, commonly geniculate at the anterior, forming a tonguelike projection that is accommodated by a low, rounded dorsal fold. The valves bear relatively numerous, rounded plications that increase in number anteriorly by bifurcation and implantation. The costae are separated by U-shaped interspaces of about the same width and amplitude. There generally are seven to nine costae that reach a size of about three-quarters of an inch in maximum width near the anterior margin on each flank of pedicle valves. Concentric ornament was not observed.

Interior of pedicle valve.—Hinge teeth are broadly set apart small oval plates, nearly parallel to the hinge line. They are unsupported by dental lamellae. Muscle scars are not impressed although there is an ill-defined, small, triangular area in the apex that apparently was the site of muscle attachment. The interior is radially grooved, reflecting the costation of the shell.

Interior of brachial valve.—The cardinalia consist of small sockets defined by strongly curved socket plates whose ventromedial edges bear small ridgelike or platelike hinge plates that are discrete and unsupported medially. The site of diductor attachment is not defined. Myophragm and adductor muscle scars are not developed. The interior is radially ridged, reflecting the costation of the shell.

Occurrence.—F fauna; UCR 5445 at Birch Creek section II–III.

Figured specimens.—USNM 171521–171524.

Subfamily SEPTATRYPINAE Kozlowski
Genus *Dubaria* Termier
Dubaria megaeroides Johnson and Boucot
(Pl. 27, figs. 13–15)

Dubaria megaeroides Johnson and Boucot, 1970, p. 267, pl. 54, figs. 10–25.

Material.—Sixty specimens with poorly preserved silicified interiors and exteriors with valves more or less welded closed.

Discussion.—Available specimens come from two beds that appear to at least locally constitute a coquina of specimens of *Dubaria* virtually to the exclusion of other brachiopods. The shells are tightly packed together and commonly flattened or deformed in some way so that their external morphology is hard to accurately characterize. In addition the interior structures are not exposed. Within these limitations and considering the best preserved of the available specimens there seems not to be any way the specimens at hand can be differentiated from *Dubaria megaeroides* from the Tor Limestone as described by Johnson and Boucot, 1970.

Occurrence.—Uppermost Pridolian; UCR 5436 and 5438 at Birch Creek section II–III.

Figured specimen.—USNM 171552.

Dubaria sp.
(Pl. 27, figs. 8–12)

Material.—Approximately forty-nine poorly preserved silicified specimens.

Discussion.—Most of the available specimens are poorly preserved fragmentary scraps but are easily characterizable. They are extraordinarily thin shelled and globose with both valves fairly strongly convex and with the brachial valve hemispherical. The pedicle valve has a long, geniculate tongue equal in width to about half the maximum width of the valves. The ventral interior is not shown on the available specimens, but the dorsal interior is characterized by a relatively small, delicate cardinalia with prominent platelike hinge plates that lie in a plane rather than converge toward the midline and which, although discrete, come relatively close to closing off the apex of the valve. There is no dorsal septum or myophragm that connects to the hinge plates basally, and its absence is probably attributable at least in part to the extremely thin-shelled nature of the specimens. The form is particularly reminiscent of *Septatrypa secreta* Kozlowski, but if the distinction between *Septatrypa* and *Dubaria* is valid (*contra* Kulkov, 1966) the similarity may be coincidental.

Occurrence.—F fauna; most specimens are from UCR 5444 and 5450 at Birch Creek section II–III and UCR 5473 near Birch Creek. Also UCR 5462, Willow Creek. Questionably identified specimens of *Dubaria* occur in UCR 5445 at Birch Creek section II–III.

Figured specimens.—USNM 171548–171551.

Genus *Cryptatrypa* Siehl
Cryptatrypa angusta n. sp.
(Pl. 26, figs. 1–18)

Diagnosis.—Small, narrowly elongate, and deeply biconvex.
Material.—Approximately 295 silicified specimens.

Exterior.—The smallness of this species seems to be characteristic, with an average length of larger specimens being approximately 5 mm. In lateral profile the valves are strongly biconvex and of about equal depth. The ventral umbo is relatively prominent and the umbo and beak protrude beyond the posterior of the brachial valve. The palintrope is narrow and ill-defined. The ventral beak is perforated by a circular foramen that appears to be mesothyrid or submesothyrid in position. The outline is generally subpyriform, but typically is modified to a narrow, elongate outline by lateral margins that tend to be subparallel. Maximum width is anterior to midlength, and valves are invariably longer than wide. Both valves tend to be deeply arched posteriorly, but the pedicle valves may develop a flattening anteriorly or even a broad, nearly imperceptible sulcation which tends to deflect the otherwise rectimarginate anterior commissure toward the brachial valve.

Radial ornament is not developed and the exteriors are smooth except for a few ill-defined concentric growth lines. These may become a little more common and prominent anteriorly on the largest specimens where growth increments are being added in such a way that the thickness of the valves is being increased at a rate greater than the length.

Interior of pedicle valve.—The teeth are small and apparently nearly circular in cross section. They are supported by ridgelike, virtually obsolescent dental lamellae that are presumed to define the muscle field, but muscle scars are not impressed. The interior of the valves is smooth.

Interior of brachial valve.—The cardinalia consist of small, divergent socket ridges and a pair of discrete hinge plates. No median septum or myophragm is developed. Positions of muscle attachment are not marked and the interior of the valves is smooth.

Comparison.—The oval, nontriangular outline separates *C. angusta* from the form Lenz (1970) called *C. nalivkini.* There is also a decided shape difference compared with *C. fahraeusi* Lenz and it must be noted that the latter species includes some specimens of *Protathyris* (Lenz, 1970, text fig. 6, 9–26). *Cryptatrypa angusta* does not resemble any of the species described by Siehl (1962). A specimen of *Atrypa canaliculata* (Tschernyschev, 1893, pl. 9, fig. 15; not Barrande) resembles *C. angusta.*

Occurrence.—F fauna; *Cryptatrypa angusta* is abundant at locality UCR 5461, Willow Creek, and is present, but less common in the stratigraphically nearby locality UCR 5462, Willow Creek. Also UCR 5450, 5451, 5453, 5454, and 5456 at Birch Creek section II–III. Also UCR 5473 at Birch Creek. Specimens of *Cryptatrypa* not identified specifically occur in UCR 5445, 5449, 5452, 5453, and 5455 at Birch Creek section II–III, at UCR 5473 at Birch Creek, and UCR 5465, west of Willow Creek; and at UCR 5469 at 1,670 ft, Pete Hanson Creek.

Discussion.—Some specimens of *Cryptatrypa* are figured on pl. 27, figs. 1–7. These are USNM 171544–171547.

Figured specimens.—USNM 171534–171543.

Genus *Atrypella* Kozlowski
Atrypella sp. E

Material.—Sixty poorly preserved silicified specimens.

Discussion.—The available material is not satisfactory on which to base a morphologic description, however, a few characterizing remarks are possible. Most of the specimens are small, lenticular, subcircular or broadly pyriform, smooth shells preserved as articulated specimens. The interiors are not well displayed. Several of the larger specimens are subcircular in outline and are slightly unequally biconvex with the brachial valve a little deeper than the pedicle valve. These specimens appear to be very slightly wider than they are long. The overall aspect is lenticular in contrast with the very deeply convex typical *Atrypella prunum* types. Interiors are not exposed except for one pedicle valve which can be determined to lack dental lamellae.

Occurrence.—Pridolian fauna; the majority of specimens discussed under this category occur in UCR 5428 at Birch Creek section II–III. The others occur in UCR 5434 at Birch Creek section II–III.

Atrypella? sp.
(Pl. 8, figs. 11–15)

Material.—Eleven silicified specimens.

Discussion.—This is rarely occurring, decidedly elongate, smooth atrypid which has not yet yielded any specimens that clearly show the interior structures; thus the identification is regarded as tentative. The outline is elongate with the lateral margins almost parallel, evenly rounded at either end, or slightly more angular posteriorly. In lateral profile the valves are unequally biconvex with a deep pedicle valve that is curved downward along its sides as well as front to back. By contrast the pedicle valve tends to be only slightly curved across the width although it is broadly curved from posterior to anterior. The whole width of the pedicle valve is involved in the dorsal deflection of the anterior commissure giving it a characteristic appearance.

Occurrence.—E fauna; this form occurs in UCR 5430 and 5431 at Birch Creek section II–III. Additional specimens that are very poorly preserved, but which may be related to this form occur in UCR 5429 at Birch Creek section II–III.

Figured specimen.—USNM 171358.

Subfamily ATRYPININAE McEwan
Genus *Atrypina* Hall and Clarke
Atrypina prosimpsoni n. sp.
(Pl. 25, figs. 12–25)

Atrypina cf. *simpsoni* Johnson 1970, p. 160, pl. 43, figs. 1–4.

Diagnosis.—Nearly lenticular, multicostate *Atrypina* with long hinge line.

Material.—Thirty-one silicified specimens from fifteen different localities.

Exterior.—The outline is subsemicircular to subtriangular, modified by a short, pointed ventral beak. In lateral profile the valves are almost plano-convex although most brachial valves are slightly convex rather than flat or concave. Pedicle valves are very gently convex for the genus. The hinge line is long and nearly straight. Cardinal angles are rounded, either acute or obtuse, but maximum width is invariably in the posterior half. The ventral beak lies close to the posterior margin without an interarea. It is apically perforate and has the foramen closed off anteriorly by a pair of small deltidial plates. The beak is nearly straight to slightly incurved. The pedicle valve may be subcarinate although no fold is separately developed. There commonly is a shallow sulcus on the brachial valve. The anterior commissure is nearly rectimarginate or may be deflected slightly ventrally.

The valves are multicostate with a prominent median pair on the pedicle valve. The median pair originates by splitting of a single broad median costa on the ventral umbo. Laterally, there commonly are three, and in some specimens four, prominent rounded costae separated by U-shaped interspaces of about the same width and amplitude. New costae are added between the primary lateral costae and the pair of median costae, as is typical of the genus *Gracianella*. The radial costae are crossed at regular intervals by prominent concentric growth lines.

Interior of pedicle valve.—The hinge teeth are platelike and suboval in outline, paralleling the nearly straight margin of the posterior valve. They are unsupported by dental lamellae and the interior is unmarked by muscle impressions or platforms. It is corrugated, reflecting the costation of the valve.

Interior of brachial valve.—The cardinalia consist of simple socket plates with anterolateral edges strongly curved to parallel the posterior margin of the valve so that the sockets are very broadly divergent. No cardinal process was observed and the notothyrial cavity is simple and open. No other internal structures are well developed. The interior is corrugated, reflecting the costation of the shell in the same degree as occurs in pedicle valves.

Comparison.—*Atrypina prosimpsoni* bears some obvious similarities to *Atrypina simpsoni* of the slightly younger *Quadrithyris* Zone (Johnson, 1970). Both species are characterized by the same outline with the broad hinge line and short ventral beak, but *Atrypina prosimpsoni* is relatively flatter and broader and it commonly has one more well-developed costa on each flank. *Atrypina prosimpsoni* also resembles the type species *Atrypina imbricata* except that the latter has coarser and fewer plications and a more rounded hinge line. Silurian species like *A. barrande, A. disparalis,* and *A. erugata* are even more coarsely plicate, narrower, and more deeply convex.

Occurrence.—F fauna; UCR 5460, 5461, 5462, 5465, and 5467 west of Willow Creek. Also UCR 5448, 5450, 5451, 5452, 5454, 5455, 5457, and 5458 at Birch Creek section II–III. Also UCR 5473 near Birch Creek.

Figured specimens.—USNM 171528–171533.

Subfamily CARINATININAE
Genus *Gracianella* Johnson and Boucot

Type species.—Gracianella lissumbra Johnson and Boucot, 1967:871, pl. 109, figs. 21–40.

Gracianella reflexa n. sp.
(Pl. 7, figs. 1–18; pl. 8, figs. 1–10)

Diagnosis.—Coarsely costate or plicate *Gracianella* with a pair of reflexed ventral plications adjacent to the midline.

Material.—A total of 653 silicified specimens.

Exterior.—The shells are suboval to pentagonal with length about equal to width; both elongate and transverse specimens occur. In larger specimens the apical angle is about 90°, but it is variable. The posterolateral margins are relatively straight to almost midlength which is the place of maximum width. Further

to the anterior the margin is subcircular or may be trigonal, completing a pentagonal outline, in shells in which a reflexed ventral pair of costae are prominent.
In lateral profile the valves are relatively lenticular and biconvex with the pedicle
valve slightly deeper than the brachial valve. The ventral beak is pointed and
straight. The delthyrium is broad and triangular and is closed by deltidial plates
except for a subhypothyrid foramen.

The pattern of radial plications of *Gracianella reflexa* is remarkable. Except
for rarely occurring secondary plications or costae the plications all are simple
and unbranched and begin at the apex of the valve. The pedicle valve has a broad,
rounded median plication, but lateral plications are angular. The lateral plications
directly adjacent to the median plication tend to flare in their width at the anterior margin of larger shells and their crests are flat or reflexed instead of curving
toward the plane of commissure. In addition to the median plication and first
lateral pair there generally are two strong and one weak lateral plications. Brachial
valves have a medial furrow and a strong elevated but narrow pair of plications
bounding the furrow. These are generally straight throughout most of their length,
but on larger specimens tend to recurve toward the midline near the anterior
commissure (pl. 7, fig. 16). It may be noted in particular with regard to the plications adjacent to the midline on the two valves that the dorsal plications are
narrow and the ventral plications are broad, especially anteriorly. In other words,
insofar as interplical furrows of one valve correspond to the plications of the
opposite valve, furrows and plications are not of equal width. It seems incorrect
to describe the valves as bearing fold and sulcus; the anterior commissure could
be designated episulcate. At a certain late growth stage the plications become
obsolete in a short distance through addition of shell material along the line of
commissure so that the interlocking nature of the plications becomes obsolescent.
In its most advanced state this type of shell secretion results in a rectimarginate
commissure having an oyster-like appearance (pl. 8, fig. 8).

Concentric ornament consists of gently imbricating growth lines over the whole
of the valves.

Interior of pedicle valve.—Hinge teeth are attached directly to the posterolateral
valve margins without dental lamellae and consist of crenulated diverging ridges.
The muscle field is relatively small, compact, and triangular. It is situated in the
apex with a tendency to emphasis of lateral lobes and with the anterior edge
elevated on a platform above a shallow cella. The interior of larger specimens is
only faintly crenulated, reflecting the plication of the shell.

Interior of brachial valve.—Sockets are narrowly triangular grooves set into
the posterolateral margins of the valve. The socket plates may be relatively prominent structures with posteroventral edges playing a prominent part in the formation of the cardinalia. Between them there generally is a broad, or even knoblike,
shell filling forming a prominent notothyrial platform. This is generally connected
anteriorly with a short myophragm that bisects poorly impressed to strongly
impressed adductor scars. Where they are present the posterior adductors are
broadly set apart lateral to the posterior portions of the anterior adductors which
lie medial, in a position on either side of the myophragm, or median ridge. The
interior may be corrugated, reflecting the plication of the shell, but is commonly
smooth owing to the deposition of secondary shell material, except for a broad

median ridge widening toward the anterior commissure and corresponding to the median furrow of the dorsal exterior.

Comparison.—This species is like no other, named or unnamed. The type species, *G. lissumbra* is transversely oval and almost smooth, so the two bear very little resemblance. In fact if the variety of morphologic shapes present in the form of various species were not known it is certain that *Gracianella reflexa* and *G. lissumbra* would be assigned to different genera. *Gracianella reflexa* differs from *G. umbra* (Barrande) in lacking a transverse-oval outline and in having much more pronounced plications. *Gracianella reflexa* differs from *G. plicumbra* and *G. crista* in the strong plication of its shell. In addition all of these species appear to have distinctive internal structures although they are variations on the same theme.

Discussion.—*Gracianella reflexa* is one of the youngest in the sequence of *Gracianella* species of the Silurian of Nevada. It lies above the great development of *G. crista, G. plicumbra,* and *G. lissumbra* of the Ludlovian, but representatives of *G. lissumbra* or a closely comparable species occur above the level of *G. reflexa.*

It seems likely that *G. reflexa* was derived from *G. plicumbra.* The closest similarities lie in that direction and among the smaller specimens of *G. reflexa* there are a few (e.g., pl. 7, figs. 17, 18), that depart from the ordinary *G. reflexa* theme in having some secondary plications that originate by bifurcation. In these the plications are of a more ordinary type and the whole shell resembles *G. plicumbra* relatively closely.

Occurrence.—E fauna; all the illustrated types are from UCR 5430 at Birch Creek section II–III. Other occurrences are in UCR 5429, 5431, 5432, and 5433 at Birch Creek section II–III.

Stauffer (1930:87) listed *"Ptychospira ferita"* from the Vaughn Gulch Limestone near Kearsarge in southeastern California. Because of morphologic similarities the form may actually be a *Gracianella reflexa.* Stauffer's material except for types, is lost and must eventually be recollected.

Figured specimens.—USNM 171343–171357.

Gracianella cryptumbra n. sp.
(Pl. 3, figs. 1–20)

Diagnosis.—A lenticular, pyriform to transversely suboval *Gracianella* with low costae.

Material.—A total of 337 silicified specimens.

Exterior.—The shells are transversely oval or pyriform in outline. In lateral profile the valves are plano-convex with only a very gently convex, subcarinate pedicle valve, which curves evenly from back to front. The ventral beak is short, straight, and pointed and the apical angle is generally large, well over 90°, although variability is considerable. Narrower specimens, with an apical angle only slightly more than 90°, are known. Maximum width is near midlength and from that point anteriorly the outline of the valves is semicircular. The carinate shape of pedicle valves corresponds to a faint sulcation of brachial valves which generally is very broad and shallow in the anterior portions.

Pedicle valves have a very broad, low median costa and an indeterminate, variable number of lateral costae, among which are secondary costae that increase by addition near the midline and also by bifurcation of lateral costae. They are of variable strength, but generally are on the low and inconspicuous side, commonly tending to obsolescence anteriorly and being variable between well-differentiated costae and barely perceptible ones. Concentric ornament occurs irregularly, most commonly at relatively widely spaced intervals. The successive growth increments generally are points of reduction in strength of the costae on specimens where these tend to obsolescence anteriorly.

Interior of pedicle valve.—The hinge teeth are broadly set apart and are simple ridges subparallel to the hinge line and set directly along the posterolateral valve margins without dental lamellae. The muscle scars are poorly impressed and are made only moderately prominent by the presence of a low, inconspicuous elevation anteriorly. The shells are nearly smooth on the interior, only faintly reflecting the external costation.

Interior of brachial valve.—The cardinalia are small and delicate and situated right along the posterior margin of the valve. They consist of a pair of very widely divergent inner socket ridges whose proximal ends meet at the apex of the valve and define a very small notothyrial cavity. There may or may not be a short myophragm extending anteriorly from the notothyrial cavity. Adductor muscle scars are not impressed. The interior is faintly to prominently corrugated, reflecting the costation of the shell.

Comparison.—*Gracianella cryptumbra* probably resembles *G. umbra* and *G. lissumbra* more than any other. It differs from *G. umbra* in being much smaller and being not quite so transversely oval. In size and outline *Gracianella cryptumbra* most closely resembles *G. lissumbra,* but that species lacks the costation that would make any comparison really close. In addition, *Gracianella cryptumbra* is gently curved from posterior to anterior in contrast to the tendency for almost a straightbacked lateral ventral outline of *G. lissumbra;* internally, these two species are very close. *Gracianella crytumbra* differs from *G. plicumbra* and *G. crista* in having much less prominent radial costae and in having many shells that are transverse rather than pyriform. *Gracianella cryptumbra* resembles *G. plicumbra* internally, but is unlike *G. crista* with its straight-edged ventral muscle platform and its ridge-bounded dorsal adductor field and central dorsal elevated knob. *Gracianella cryptumbra* differs from *G. reflexa* on all major points of shell shape, costation, and internal structures.

Occurrence.—Pridolian fauna; all the illustrated types are from UCR 5428 at Birch Creek section II–III. A few poorly preserved specimens from UCR 5431 and 5437 at Birch Creek section II–III are assigned to *G. cryptumbra.*

Figured specimens.—USNM 171298–171309.

Occurrence.—E fauna; UCR 5434, UCR 5436, and UCR 5437 at Birch Creek section II–III.

Genus *Sibirispira* Alekseeva, 1968

Type species.—Sibirispira inflata Alekseeva, 1968:200, text figs. 1, 2.

Sibirispira? sp.
(Pl. 25, figs. 8–11)

Occurrence.—F fauna; UCR 5444 at Birch Creek section II–III.
Material.—Fifty-three silicified specimens.

Discussion.—These are small subcircular, lenticular, almost flat shells that resemble *Gracianella* except that they are multicostate.

Figured specimens.—USNM 171525–171527.

Family LISSATRYPIDAE Twenhofel
Genus *Lissatrypa* Twenhofel
Lissatrypa sp.
(Pl. 27, figs. 16–21)

Material.—Fifteen silicified specimens.

Exterior.—The shells are suboval and very slightly transverse to nearly circular in outline and are subequally biconvex, lenticular, in lateral profile. Some specimens have a pedicle valve slightly deeper than the brachial valve. The hinge line is only gently curved but is short, combining with strongly rounded posterolateral margins to give an overall rounded outline. The ventral beak is very short and strongly incurved and is appressed, without an interarea, to the dorsal umbo. Smaller specimens have a relatively more prominent hinge line and less rounded posterolateral margins giving the valve a shouldered appearance. Maximum width is posterior of midlength and the valves are evenly semicircular from the place of maximum width around the anterior margin.

There is no radial ornament. The concentric ornament consists of few prominent growth lines developed in an irregular fashion near the anterior margin of some larger individuals. The exterior of most specimens is very rough, indicating the presence of very abundant spinules like those developed on the genus *Nucleospira*. It is not generally known that the lissatrypid brachiopods have this kind of spinule development, but it has been seen excellently preserved on a collection of silicified specimens of *Lissatrypoidea* (or *Nanospira?*) from the Brownsport Formation of western Tennessee (USNM 12258).

Interior of pedicle valve.—The hinge teeth are triangular protuberances set along the valve margin and separated from the posterior margin of the pedicle valve by a pair of transverse grooves that receive the posterior edge of the brachial valve. Dental lamellae are absent. The muscle field is bilobate and impressed. It is divided medially by a V-shaped myophragm that is elevated anteriorly into a small triangular platform. The interior is smooth.

Interior of brachial valve.—The sockets lie incised adjacent to the posterior valve margin and are defined medially by strong socket plates that extend ventrally above the level of the commissure. The area between the socket plates is filled in, forming a solid mass that is slightly knoblike medially. The complete structure is commonly described as a cardinalia with inner hinge plates or the medial filling may be called a cardinal block after Siehl (1962:185). The available material is inadequate to show the details of the structure. The base of the cardinal block is joined to a low, ridgelike myophragm that divides paired adductor impressions. The posterior pair is small and nearly circular compared with a much larger and broader anterior pair which is best impressed laterally and posteriorly and which blends with the shell interior anteriorly. The remainder of the interior is smooth.

Occurrence.—Basal F fauna; UCR 5441 and 5442 at Birch Creek section II–III. It must be pointed out that both of these occurrences have stratigraphic positions that must be questioned. One is of float material and the other is questionably assigned to a horizon by resemblance of the fauna to another that is well placed in the measured sequence.

Figured specimens.—USNM 171553–171556.

Suborder RETZIOIDEA

Discussion.—Only a single retzioid genus occurs in the Roberts Mountains Formation and that is *Rhynchospirina* in the F fauna of Gedinnian age.

Superfamily RETZIACEA Waagen
Family RHYNCHOSPIRINIDAE Schuchert and LeVene
Genus *Rhynchospirina* Schuchert and LeVene
Rhynchospirina siemiradzkii Kozlowski
(Pl. 29, figs. 1–17)

Rhynchospirina siemiradzkii Kozlowski, 1929:214, text figs. 76, 77; pl. 9, figs. 27–34.

Material.—A total of 107 silicified specimens from eleven different localities.

Exterior.—The shells are pyriform in outline and subequally biconvex in lateral profile. Both valves are relatively strongly convex so that the articulated specimens are well rounded rather than lenticular. The hinge line is short and rounded and maximum width is about at midlength. The beak is incurved to the nearly erect position and is perforated by a circular foramen of mesothyrid position. The apical angle is a little less than 90° and may remain unchanged because of nearly straight posterolateral margins, or may increase slightly as the shell widens. The outline is nearly semicircular. Both valves are slightly sulcate medially and most specimens have a smaller pair of costae in the sulcus, originating from the umbo rather than the beak. The valves are covered by strong, elevated, rounded costae separated by relatively deep, narrow, U-shaped interspaces. The costae do not increase in number anteriorly by splitting or by intercalation. They are crossed, on some specimens, by a few inconspicuous concentric growth lines.

Interior of pedicle valve.—There are no dental lamellae. The hinge teeth are small and bluntly rounded. The interiors do not have a muscle scar impressed but are radially corrugated, reflecting the costation of the shell.

Interior of brachial valve.—The cardinalia consist of a posteriorly curved cardinal plate which is quadrate in outline and parallels the plane of commissure. Its anterior edge is cleft by a small V-shaped indentation lateral to which there is a pair of lobes that must have served as crural bases. The whole structure is supported by a short bladelike median septum extending about a fourth of the length of the valve. Muscle scars are not impressed and the interior is smooth except for radial corrugations, reflecting the costation of the shell.

Occurrence.—F fauna; UCR 5459, 5460, 5461, 5463, and 5465 at Willow Creek. Also UCR 5450, 5453, 5454, 5456, and 5458 at Birch Creek section II–III. Also UCR 5469 at 1,670 ft, Pete Hanson Creek.

Figured specimens.—USNM 171563–171569.

Suborder ATHYRIDOIDEA

Discussion.—The athyridid brachiopods are common in beds of Pridolian and Gedinnian age in the Roberts Mountains, but are represented by conservative and unappealing stocks. *Nucleospira* and *Meristina* are the common forms; rare *Meristella* occurs in the upper F fauna. A species of *Protathyris* is notable in the E fauna of Pridolian age.

Superfamily ATHYRIDACEA M'Coy
Family NUCLEOSPIRIDAE Davidson
Genus *Nucleospira* Hall

Discussion.—F fauna specimens are larger than specimens of *Nucleospira* from Silurian faunas of the Roberts Mountains Formation.

Nucleospira spp.
(Pl. 28, figs. 10–15)

Occurrence.—UCR 5460, 5461, 5462, 5464, 5465, 5466, and 5467, at Willow Creek. Also UCR 5428, 5444, 5445, 5446, 5448, 5449, 5450, 5455, 5457 and 5458, all at Birch Creek section II–III. Also at UCR 5473 near Birch Creek. Also UCR 5469 at 1,670 ft and UCR 5470 at approximately 1,870 ft, both at Pete Hanson Creek.

Figured specimens.—USNM 171561, 171562.

Family ATHYRIDIDAE M'Coy
Subfamily PROTATHYRIDINAE Boucot, Johnson, Staton
Genus *Protathyris* Kozlowski
Protathyris sp. E
(Pl. 8, figs. 16–25)

Discussion.—This species does not closely resemble any of those described by Kozlowski (1929), and other species that can confidently be assigned to *Protathyris* are poorly known. It should be noted that this species lacks mystrochial plates which would indicate an assignment to *Didymothyris* Rubel and Modzalevskaya (1967) instead of *Protathyris*.

Material.—Sixty silicified specimens.

Exterior.—The shells vary from subcircular to elongate suboval in outline and are subequally biconvex in lateral profile. The hinge line is short and curved and the maximum width commonly is slightly posterior of midlength. The ventral beak is only short and slightly incurved, to the suberect position. The delthyrium is triangular and open and the foramen is submesothyrid. Neither fold nor sulcus is developed and the exterior is smooth.

Interior of pedicle valve.—There is a pair of short, platelike dental lamellae that diverge slightly and which are subparallel in cross section. The muscle field is not impressed and the interior is smooth.

Interior of brachial valve.—The sockets are short but deep, adjoining the posterolateral edge of the valve and attached on their medial edges to a cardinal plate with a depressed median trough. The latter uncommonly is filled in with a ridgelike deposit (pl. 8, fig. 25). The cardinal plate is unsupported except by the socket plates. Beneath it the adductor tracks, when impressed, compose an elongate lanceolate pair. The remainder of the interior is smooth.

Occurrence.—UCR 5430 at Birch Creek section II–III.

Figured specimens.—USNM 171359–171362.

Family MERISTELLIDAE Waagen
Subfamily MERISTELLINAE Waagen
Genus *Meristella* Hall
Meristella cf. *wisniowskii* Kozlowski
(Pl. 28, figs. 1–9)

Meristella wisniowskii Kozlowski, 1929:219, text fig. 80; pl. 11, figs. 36–41.
Meristella wisniowski Nikiforova, 1954:153, pl. 17, figs. 3, 4.

Material.—Twenty-one silicified specimens.

Discussion.—The genus *Meristina* Hall is the common meristellid of the Roberts Mountains Formation, but a few occurrences high in the Gedinnian yield specimens that belong to *Meristella*. None of the small number of specimens available is well preserved and only a few display internal characteristics that can be confidently ascribed to *Meristella*. Those few have relatively deeply impressed and broad ventral muscle impressions without dental lamellae, but this can only be seen on larger specimens. Smaller ones, without thick shell deposits, have thin dental lamellae and nonimpressed muscle fields, and are inseparable from *Meristina*.

Brachial valves bear a septalium-like cardinalia supported by a median septum as is typical. The septalium variably is open, filled by a deposit of shell material, or split by the posterior extension of the median septum.

Occurrence.—F fauna; UCR 5461 at Willow Creek. Also UCR 5457 and 5458 at Birch Creek section II–III. Also UCR 5469 at Pete Hanson Creek. Of the occurrences listed above, only a few specimens from UCR 5457 display the critical structures of the ventral interior.

Figured specimens.—USNM 171557–171562.

Suborder SPIRIFEROIDEA

Discussion.—The delthyrid spirifers are represented by *Howellella* in the Pridolian and Gedinnian, but although the genus occurs in many collections it occurs in small numbers, represented by small specimens of a conservative sort. The genus *Delthyris* occurs in Ludlovian Pridolian age collections, but has no representation in the Devonian in Nevada. Particularly significant is the occurrence of *Megakozlowskiella* in the F fauna of Gedinnian age where it is represented by a form probably related to the Bohemian species *M.? inflectens*. If the Bohemian species is correctly assigned to *Megakozlowskiella* then that genus has a rather wide distribution in the Gedinnian as is true of the Siegenian (Boucot, Johnson, and Talent, 1969).

The reticulariid spirifers are represented by two genera. The first is *Tenellodermis*, Havlíček, 1971, a new genus known to range through Pridolian and Gedinnian age beds in the Old World. Its Nevada occurrence is within a single fauna, for example, the E fauna of Pridolian age where it seems to be particularly diagnostic. It occurs elsewhere in Nevada in the company of *Dubaria megaeroides* in the Tor Limestone (Johnson and Boucot, 1970). The second reticulariid genus is *Undispirifer*, represented by *U.* cf. *laeviplicatus* in the F fauna of Gedinnian age. It occurs there in several collections, but is uncommon.

The ambocoeliid spirifers are represented by *Metaplasia* in the Gedinnian. This same fossil is also known in beds as old as Pridolian in Yukon Territory (Lenz, 1970). In Nevada no *Metaplasia* has been found in beds older than Gedinnian and the rare Pridolian fauna occurrence is exemplified by *Alaskospira?*.

Finally, the important post-Silurian genus *Cyrtina* has a good representation in the F fauna of Gedinnian age in the Roberts Mountains, and is relatively common in a number of collections.

Superfamily DELTHYRIDACEA Phillips
Family DELTHYRIDIDAE Phillips
Subfamily DELTHYRIDINAE Phillips
Genus *Howellella* Kozlowski
Howellella spp.
(Pl. 31, figs. 1, 2)

Material.—A total of 216 silicified specimens.

Discussion.—Although of worldwide distribution and common occurrence the genus *Howellella* is not a common fossil in the Roberts Mountains Formation. The specimens that do occur are few in number and are scattered among numerous collections and they invariably are poorly preserved. The material from the Pridolian-Gedinnian interval, examined here, defies description of any consequence, largely because of small size and poor preservation and because of some variability. *Howellella* species recognized here are commonly relatively strongly biconvex, thin-shelled brachiopods with a variable development of plications. The internal structures are small and delicate as a rule, but in one collection (UCR 5457) a distinct species is represented mostly by fragments of pedicle valves that show the development of long and thick dental lamellae of the type that Kozlowski (1929, pl. 10, figs. 18, 19) illustrated in specimens of *Howellella angustiplicatus*.

Occurrence.—E and F faunas; UCR 5461 and 5465 west of Willow Creek. Also, UCR 5428, 5433, 5437, 5444, 5445, 5447, 5448, 5450, 5451, 5452, 5453, 5454, 5455, 5457, 5458 at Birch Creek section II–III. Also UCR 5469 at 1,670 ft, Pete Hanson Creek section.

Figured specimen.—USNM 171578.

Genus DELTHYRIS Dalman
Delthyris spp.
(Pl. 3, figs. 21–26)

Material.—Twenty-seven silicified specimens from two localities.

Exterior.—The more common form is a small, transversely oval species of unequally biconvex lateral profile. The pedicle valve is one and a half to two times as deep as the brachial valve and bears a short hinge line and rounded cardinal angles. The interarea is small and triangular and is only relatively low and apsacline. The beak is slightly incurved, stubby, and pointed. The plications are low and rounded, commonly including about three on each flank of a pedicle valve lateral to a shallow ventral sulcus. The dorsal fold is low and rounded, but slightly carinate and its anterior commissure is indented medially, forming a V at the midline that accommodates the pointed anterior tip of a ventral sulcal tongue. Fine ornament is not preserved.

Interior of pedicle valve.—Hinge teeth are not well preserved, but evidently were small and delicate. The dental lamellae are thin and platelike, restricted to the posterior; they descend to the base of the valve in a subparallel fashion adjacent to a thin, bladelike median septum in the posterior portion of the valve. The shell material is thin and corrugated, reflecting the plication of the shell.

Interior of brachial valve.—The cardinalia are very delicate, consisting of inclined, triangular hinge plates lacking strong crural plate support. The shell material is thin and corrugated reflecting the plication of the shell.

Occurrence.—Pridolian and E faunas; UCR 5428, and 5430 at Birch Creek section II–III. The latter includes only three specimens, which are a little larger and more faintly plicate than the ones from the lower horizon. They therefore are likely a different species.

Figured specimens.—USNM 171310–171313.

Subfamily KOZLOWSKIELLININAE Boucot
Genus *Kozlowskiellina* Boucot
Subgenus *Megakozlowskiella* Boucot
Megakozlowskiella cf. *M.? inflectens* (Barrande)
(Pl. 30, figs. 1–9)

Spirifer inflectens Barrande, 1879, pl. 2, figs. 9–11.
Cyrtinopsis inflectens Havlíček, 1959a:144, text figs. 65–67; pl. 23, figs. 5–7.

Discussion.—There is a distinct resemblance of this species of *Megakozlowskiella* to "*Spirifer*" *inflectens* Barrande which Havlíček (1959a) assigned to the genus *Cyrtinopsis*. For example, compare the dorsal and posterior views of the brachial valve in plate 30, figures 3, 4, with the same views of Havlíček (1959a, pl. 23, figs. 5, 7). Neither the fine ornament nor the cardinalia of "*Spirifer*" *inflectens* are well known, and, because *Cyrtinopsis* is typically a Middle Devonian genus, the assignment of "*Spirifer*" *inflectens* to it seems doubtful. Part of the basis for Havlíček's assignment of "*Spirifer*" *inflectens* to *Cyrtinopsis* rests on the fact that the long dental plates converge and join the sides of the median septum just as they do in *Cyrtinopsis,* as illustrated by Boucot (1957, pl. 1). Typically, *Megakozlowskiella* has stubby pseudodental plates that hang free from the inner edges of the delthyrium and join neither the base of the valve nor the median septum. This, however, is not invariably the case and other specimens from Nevada, particularly those figured by Johnson (1970, pl. 72, fig. 23) from the *Quadrithyris* Zone, very definitely do have dental plates that join the median septum as in "*Spirifer*" *inflectens*. So do some specimens of *Megakozlowskiella* from the *Acrospirifer kobehana* Zone in the lower Emsian of central Nevada. Considering the evidence presently available it may very well be more likely that "*Spirifer*" *inflectens* belongs to *Megakozlowskiella* than to *Cyrtinopsis* where it was assigned by Havlíček, and if so the available specimens may be closely related or conspecific with "*S.*" *inflectens*. Unfortunately, it has never been possible for the writers to obtain specimens of "*Spirifer*" *inflectens* in order to more adequately assess its morphology. It may not be possible to determine the fine external ornament because the specimens are from hard limestone and the exterior shell is generally spalled off. A specimen probably could be prepared as an internal mold showing the cardinalia. The cardinalia of *Megakozlowskiella* and *Cyrtinopsis* are quite different as shown by Boucot (1962, pl. 52). Hopefully, the true assignment of *"Spirifer" inflectens* will eventually be determined.

Another comparable species is *Kozlowskiellina paradoxa* Kulkov (1963, pl. 8) from the Siegenian of the Altai Mountains.

Material.—Five silicified specimens from three localities.

Exterior.—The valves are broadly transverse, almost twice as wide as long with a long straight hinge line and a modified subsemicircular to subtriangular outline. In lateral profile the valves are unequally biconvex with the pedicle valve about three times as deep as the brachial valve. The interarea is well formed, triangular, slightly curved, and apsacline, and it is cleft medially by an open triangular delthyrium.

The valves are strongly plicate with rounded, elevated plications, separated by deep U-shaped interspaces. There are four plications on each flank of a brachial valve lateral to a broad, low, rounded fold. The pedicle valve bears a median sulcus that becomes relatively broad and flat-bottomed anteriorly. The concentric ornament consists of numerous well-developed, closely spaced, frilly growth lines. The fine ornament was not observed because of rough preservation of exteriors.

Interior of pedicle valve.—Long, thin, platelike, subparallel to slightly convergent dental lamellae are present attached to the sides of a thin bladelike median septum in the posterior portion of the valve. In the one available well-preserved ventral interior (pl. 30, fig. 5) it appears that a bladelike median septum and discrete dental lamellae were formed and later joined. The septum is thickened by secondary shell material which is continous from one side of the septum, across its dorsal edge, to the other side of the septum. Muscle scars are not impressed, but the interior is deeply corrugated, deflecting the plication of the shell.

Interior of brachial valve.—The sockets are composed of divergent, curved socket plates bounding the notothyrial cavity which bears a bilobate cardinal process. No muscle scars are impressed, but the interior is deeply corrugated, reflecting the plication of the shell.

Occurrence.—F fauna; UCR 5466 west of Willow Creek; also UCR 5455 at Birch Creek section II–III, and UCR 5469 at 1,670 ft, Pete Hanson Creek. One figured specimen is from the Pete Hanson Creek locality UCR 4449, a collection not otherwise studied as a part of this report.

Figured specimens.—USNM 171570–171572.

Family RETICULARIIDAE Waagen
Genus Tenellodermis Havlíček

Type species.—*Tenellodermis microdermis* Havlíček, 1971:12.

Diagnosis.—Small aseptate Reticulariidae of transverse outline, stubby insignificant ventral beak, and a nearly cataclise, flat ventral interarea. Deltidium hoodlike. Plications broad, low, variable in size and few in number, commonly simple, but splitting occurs. Internally, dental lamellae are high but very short; crural plates are lacking.

Discussion.—The genus *Tenellodermis* was proposed for a small group of reticulariid species that occur in the Ludlovian, Pridolian, and Gedinnian. Species assigned, in addition to the type species, are *Spirifer tenellus* Barrande (1848, p.

161, pl. 18, fig. 3) and *Spirifer (Crispella?) orphanus* Kozlowski (1929, p. 198, pl. 9, fig. 35), and *T. matrix* n. sp.

Comparison.—Tenellodermis is obviously distinct from septate genera and it need not be compared with large, completely smooth reticulariid genera either. This leaves a residue of faintly to markedly plicate reticulariid genera including *Elytha, Elythyna, Undispirifer,* and *Pinguispirifer. Elytha* is larger, has a rounded outline and double-barreled spines. Internally, it differs in having long divergent dental lamellae. *Elythyna* also is larger, with a rounded outline and has longer dental lamellae, a somewhat impressed ventral musculature, and crural plates. *Undispirifer* is larger and has a rounded outline; it also differs in having long, divergent dental lamellae. *Pinguispirifer,* thought to be an eospiriferid by Havlíček (1959a), was included in the Reticulariidae by Boucot (1962) and there is some similarity because of the short dental lamellae of *Pinguispirifer.* It differs in being larger and in having a different valve form, but the similarities are close enough that it might be suggested that *Tenellodermis* is the ancestor of *Pinguispirifer.*

Tenellodermis matrix n. sp.
(Pl. 9, figs. 1–19)

"Howellella" cf. *tenella* Johnson and Boucot, 1970:268, pl. 54, figs. 1–9; not Barrande, 1848.

*Material.—*A total of 250 silicified specimens.

*Exterior.—*Pedicle valves are subtriangular to roughly rhomboidal in outline. Brachial valves are subtriangular to transversely oval. In lateral profile the valves are unequally biconvex, with a deep pedicle valve and a gently convex brachial valve. The ventral palintrope is of moderate height with the beak incurved over an apsacline, triangular interarea. The delthyrium is high, triangular, and narrow, enclosing an angle slightly less than 45°. It has a very characteristic deltidial structure, consisting of a pair of strong plates that lie adjacent to the delthyrium and project normal to the interarea along the whole length of the delthyrium. Between these plates a single strong triangular plate closes off approximately the apical two-fifths of the delthyrium. The dorsal interarea is poorly developed and nearly linear.

There is a shallow U-shaped sulcus of somewhat variable strength on the pedicle valve and a rounded fold on the brachial valve. These medial structures are somewhat accentuated by the presence of sulcus-bounding ridges on the pedicle valve and by a furrow on either side of the dorsal fold. In addition there may or may not be one or two well-rounded plications on each flank of each valve. These lateral plications are separated by broad, shallow, U-shaped interspaces and their development is evanescent. Whether plications on the flanks are well marked or not, ridges and furrows bounding the medial structures are almost invariably well developed. On some specimens there is a gentle median furrow developed on the dorsal fold, mirroring the development of a lobation of the ventral sulcus.

The external ornament consists of fine radial striae on the posterior parts of the valves, suggesting an eospiriferid type ornament, but in the anterior half of larger specimens there is a common development of closely spaced, lamellose, concentric growth lines. These betray the very gentle lobation of the sulcal tongue on some specimens and on others there is an uncommon tendency for the development of branching of the plications.

Interior of pedicle valve.—Hinge teeth are delicate points at the medial dorsal edges of the delthyrium. They are supported basally by a pair of high but anteriorly short dental lamellae, almost subparallel to the base of the valve where they join, about at the position of the grooves which correspond externally to ridges bounding the ventral sulcus. The dental lamallae are high but short and diverge only very slightly anteriorly (pl. 9, fig. 12). Between the dental lamellae and the floor of the valve there is a relatively long, sharp, ridgelike myophragm that extends anterior to the distal edges of the dental lamellae. The shell material is very thin and the interior surface is corrugated, reflecting the degree of shell plication.

Interior of brachial valve.—The sockets are widely divergent, shallow, narrow, cylindroidal grooves defined by socket plates that are elevated, without support, well above the floor of the valve medially. The divergent inner edges of the socket plates are attached to well-developed, triangular crural bases that converge toward the base of the valve and toward the midline, but which are not connected to the floor of the valve by crural plates. Instead, the posterior ends of the crural bases are attached to a ridgelike thickening in the apex. The site of diductor attachment is a bilobed, angular protuberance filling the notothyrium. Adductor muscle scars are not impressed and, because the shell material is very thin, the interior is corrugated, reflecting the plication of the shell.

Comparison.—*Tenellodermis matrix* differs from *T. orphanus* (Kozlowski, 1929, pl. 9) in having a deeper ventral sulcus and a more prominent dorsally extending tongue. In addition, the Nevada shells have a slightly stronger tendency toward a triangular outline compared with the trapezoidal outline of *T. orphanus*. *Tenellodermis matrix* differs from *T. tenella* (Barrande; Havlíček, 1959a, pl. 28) in almost identical ways as it does from *T. orphanus*. Primarily, the depth and narrowness of the ventral sulcus of *T. matrix* distinguishes it from the slightly wider and less deep ventral sulcus of *T. tenella*.

Discussion.—Elsewhere the writers (Johnson and Boucot, 1970) have reported this species from the Tor Limestone of the Toquima Range of central Nevada and suggested a Pridolian age for the fossil-bearing horizon. This is borne out in the present work by the position of this fossil in the Birch Creek section, lying as it does above the well-identified Ludlovian occurrences and above the highest pentamerids, but below fossils indicative of the Gedinnian, or earliest Lower Devonian.

Occurrence.—E fauna; UCR 5429, 5430, 5431, and 5434, at Birch Creek section II–II.

Figured specimens.—USNM 171363–171370.

Genus *Undispirifer* Havlíček

Discussion.—The type species of *Undispirifer* is from the Middle Devonian of Germany, but the genus has never been studied outside its type region and its range cannot be regarded as fixed to the Middle Devonian. On the contrary, the specimens studied here, from beds of Gedinnian age, show all the morphologic characters, internal and external, which appear to be diagnostic for the genus. These are a rounded outline and biconvex development of the valves with faint to well-marked radial plications crossed by concentric growth lines whose anterior edges bear rows of small, radially aligned spine bases. Internally, the dental lamellae are long and divergent, unobscured by thick deposits of secondary shell material. The cardinalia lack strongly developed crural plates; they may be variably developed or absent.

Discrimination of spiriferids belonging to the families Delthyrididae and Reticulariidae in the Silurian and in the Early Devonian is a difficult matter. The variable strength and common obsolescence of plications seems to be the hallmark of the early reticulariids. No distinctive fine ornament is present in these early reticulariid brachiopods that unequivocally distinguishes them from members of the Delthyrididae. There is in some specimens, however, a notable departure from the delthyrid ornament style, exemplified by the development of concentric lamellae that bear spine bases only on their anterior edges and which are not continuous with radially aligned ridges that cross the whole length of the numerous lamellae.

The species described below, assigned to *Undispirifer,* is not as disconnected from the ordinary Middle Devonian occurrences of the genus as one might suspect because the authors have seen similar shells in both Gedinnian and Siegenian age beds from the Canadian Arctic Islands.

Undispirifer cf. *laeviplicatus* (Kozlowski)
(Pl. 30, figs. 10–18)

Spirifer (Crispella) laeviplicatus Kozlowski, 1929:195, pl. 10, figs. 22–27.
Spirifer (Howellella) laeviplicatus Nikiforova, 1954:144, pl. 16, figs. 6–8.

Discussion.—It is believed that Kozlowski's species *laeviplicatus* is not a proper member of *Howellella.* Instead it seems more likely to be a reticulariid because of the ornament illustrated by Kozlowski (1929, pl. 10, fig. 27). In addition there are some differences at the specific level. The specimens from central Nevada appear to be much smaller on the average and are thin-shelled. Furthermore, they may be somewhat more strongly ribbed when compared with the specimens illustrated by Kozlowski (1929, pl. 10). The Nevada specimens, however, do agree rather closely in the strength of the development of plications with specimens illustrated by Nikiforova (1954, pl. 16, figs. 6, 7). Although the specimens illustrated by Kozlowski and Nikiforova therefore differ somewhat in this regard, the difference is very likely attributable to intraspecific variation. Similar variation in the strength of plications is exhibited in the Nevada specimens.

Material.—About 370 silicified specimens.

Exterior.—The shells are generally transverse-oval in outline and unequally biconvex in lateral profile with the pedicle valve two to three times as deep as the brachial valve and with a relatively prominent, curved ventral umbo and beak. The ventral interarea is low, triangular, and apsacline. It is relatively narrow, occupying about two-thirds of the maximum width of the valves. The delthyrium is open and triangular, including an angle of about 60° or a little less and is bounded by ridgelike deltidial plates projecting posteriorly. The dorsal interarea is nearly linear and poorly defined. The pedicle valve bears a shallow, broad, U-shaped sulcus and the brachial valve bears a corresponding broad, low fold.

Plications are developed on the flanks of both valves and are generally poorly defined. On specimens in which plications are best developed there are as many as four on the flank of a pedicle valve and commonly two or three on each flank of brachial valves. A few concentric growth lines commonly may be seen in the anterior portions of larger specimens, but they are poorly defined. Poor preservation precludes the possibility of observing any fine radial ornament, if it was present.

Interior of pedicle valve.—The hinge teeth are rather small and pointed, and are supported by dental lamellae that first converge for a short distance and then bend to diverge through most of their distance toward the base of the pedicle valve. They diverge slightly from the midline as they are traced anteriorly. The dental lamellae are relatively thin and delicate and are not accompanied by umbonal thickening of any significance. Ventral muscle scars are not impressed. No median septum or myophragm is developed. The interior of the shell is faintly corrugated, reflecting the degree of plication of the shell.

Interior of brachial valve.—The cardinalia are relatively delicate and consist of thin platelike crural bases adjoining narrow, anteriorly divergent sockets. The crural bases are not connected with the base of the valve via the introduction of crural plates. The site of diductor attachment surely was very small and is poorly preserved so that the nature of the myophore is not determinable. Muscle scars are not impressed and the interior is gently corrugated, reflecting the degree of plication of the shell.

Occurrence.—F fauna; the largest number of specimens is from locality UCR 5461. Additional specimens are in collections UCR 5460, 5462, and 5465 west of Willow Creek. Also UCR 5443 and UCR 5449 at Birch Creek section II–III.

Figured specimens.—USNM 171573–171577.

Family AMBOCOELIDAE George
Genus METAPLASIA Hall and Clarke
Metaplasia lenzi n. sp.
(Pl. 31, figs. 20–28)

Metaplasia sp. Lenz, 1970:497, pl. 87, figs. 5–15.

Diagnosis.—*Metaplasia* with a flat dorsal fold and bounding furrows.

Material.—A total of 145 silicified specimens from eleven different localities.

Discussion.—There may possibly be two species of *Metaplasia* in the Gedinnian beds of the Roberts Mountains Formation. The best preserved is a species with a fold on each valve, but a few specimens have been seen with a ventral median furrow. One specimen of the latter type is illustrated (pl. 31, figs. 16–19), but of it no more will be said.

Exterior.—The valves are transverse in outline and roughly triangular. In lateral profile they are strongly unequally biconvex with a nearly flat brachial valve and a deep pedicle valve having a long ventral palintrope. The latter is moderately curved with a triangular, apsacline interarea approaching the cataline position. The interarea is cleft by an open triangular delthyrium enclosing about 60°. The hinge line is relatively long and straight but the cardinal angles are rounded and blunt. Maximum width is posterior to midlength. Pedicle valves are somewhat carinate with a slightly flattened median sector set off by a pair of radial furrows laterally. The combination of features forms what is essentially a ventral fold. The brachial valve bears a nearly flat, low, median fold bounded by a pair of lateral furrows, but in some cases the fold may be somewhat depressed between the bounding furrows to form a sulcus-like structure. Because the median structure is low and variable there can be no one term that consistently describes the deflection of the anterior commissure.

There is no radial ornament. The concentric ornament is rarely developed and consists of a few widely spaced growth lines.

Interior of pedicle valve.—Hinge teeth are small and poorly preserved. Dental lamellae are absent. The interior is smooth, without muscle impressions or other markings.

Interior of brachial valve.—The cardinalia consist of divergent cylindroidal sockets supported by crural plates that are widely spaced, but in one shell that has crural plates developed most strongly they are convergent toward the midline, forming a cruralium-like structure. The cardinal process appears to be a simple knoblike structure. The interior is smooth.

Comparison.—*Metaplasia pyxidata* has a more triangular outline; *M. paucicostata* has a more flared dorsal fold; *M. minuta* has a strong median groove on the pedicle valve.

Occurrence.—F fauna; UCR 5460, 5461, 5462, and 5465, west of Willow Creek. In addition there are specimens in UCR 5446, 5448, 5450, 5456, and 5458 at Birch Creek section II–III. In addition there are specimens in UCR 5473 in the Birch Creek area and UCR 5469 at 1,670 ft, Pete Hanson Creek.

Figured specimens.—USNM 171585–171590.

Genus *Alaskospira* Kirk and Amsden

Alaskospira? sp.

Discussion.—A single collection from the Pridolian interval has yielded eight poorly preserved silicified specimens that appear to belong to *Alaskospira*. They are too fragmentary and poorly preserved on which to base a description, but are worth mentioning because they represent the highest probable occurrence of the genus. The specimens are from Pridolian fauna locality UCR 5428 at Birch Creek section II–III.

Superfamily CYRTINACEA Frederiks
(*nom. transl.* Johnson, 1966)
Family CYRTINIDAE Frederiks
Genus *Cyrtina* Davidson
Cyrtina sp.
(Pl. 31, figs. 3–15)

Material.—A total of 199 silicified specimens.

Exterior.—This is a small, broadly transverse, subtrigonal species with a strongly unequal, biconvex, lateral profile. The brachial valve is very gently convex and the pedicle valve is very deep because of the development of a long palintrope. The ventral interarea is high, triangular, flat, and catacline. It is cleft medially by a narrow triangular delthyrium. There is no well-developed dorsal interarea. The pedicle valve is subpyramidal with only slightly curved lateral slopes. It bears a narrow, V-shaped sulcus and the brachial valve bears a low, rounded, flat-topped fold. Both valves bear two, three, or four rounded plications separated by relatively deep U-shaped interspaces on each flank. The concentric ornament is commonly lamellose and may be accentuated anteriorly.

Interior of pedicle valves.—Hinge teeth are small and pointed. They lie along ridgelike tracks inside the edges of the delthyrium and join medially with a tube-like tichorhinum which has a length of approximately half that of the palintrope. It may be seen from the exterior to close approximately half the height of the delthyrium. The tichorhinum is supported basally by short, bladelike median septum.

Interior of brachial valve.—The cardinalia are poorly preserved on the available specimens, but appear generally to consist of widely separated sockets incised into the posterior shell margin and adjoined medially by triangular, platelike crural bases and an irregular moundlike cardinal process. Muscle scars are not impressed and the interior is corrugated by the radial plications of the shell.

Occurrence.—F fauna; UCR 5459, 5460, 5461, 5462, 5465, west of Willow Creek. Also, UCR 5445, 5449, 5446, 5447, 5450, 5451, 5452, 5453, 5454, and 5455 at Birch Creek section II–III. Also UCR 5473 near Birch Creek and UCR 5469 at 1,670 ft, Pete Hanson Creek.

As noted above *Cyrtina* sp. occurs in numerous collections in the Gedinnian of Nevada, but it generally is not abundant. Only three of the collections listed above have *Cyrtina* sp. as a common element. These are UCR 5461, 5462, and 5469. All these collections have thirty-five or more specimens. The lowest occurrence is listed in E fauna locality UCR 5431 where *Cyrtina* sp. is represented by a single specimen. It is a small silicified individual with pedicle and brachial valves still articulated and because of this it could have floated and entered that collection during processing. The lowest occurrence of *Cyrtina* certainly needs verification. The next higher occurrence is that of UCR 5445 which includes six specimens of *Cyrtina* sp. so that the presence of the genus at that level need not be doubted.

Figured specimens.—USNM 171579–171583.

APPENDIX OF LOCALITIES

For localities not plotted on measured sections full details will be provided in a summary paper, in preparation. Localities have been entered in both the UCR system (University of California, Riverside, Department of Geological Sciences) and in the USNM system (National Museum of Natural History, Smithsonian Institution). The correspondence of UCR numbers and USNM numbers is listed below.

LOCALITY NUMBERS

UCR	USNM	UCR	USNM
5428	12746	5452	12762
5429	12747	5453	12763
5430	13226	5454	12764
5431	12748	5455	12765
5432	12749	5456	12766
5433	12745	5457	12327, 12767
5434	13227	5458	12768, 13231
5436	12750	5459	12328
5437	13228	5460	12329
5438	12751	5461	12330
5439	12753	5462	12331
5440	12754	5463	13209
5441	12752	5464	13210
5442	12755	5465	13211
5443	12755	5466	13212
5444	13229	5467	13213
5445	13230	5468	13224
5446	12756	5469	13250
5447	12757	5470	13251
5448	12758	5473	13238
5449	12759	——	10794
5450	12760	——	12865
5451	12761	——	10795

LITERATURE CITED

ALBERTI, G. K. B., W. HAAS, and A. R. ORMISTON
  1971. Discovery of *Warburgella rugulosa* (Alth, 1874) in Gedinnian strata of central Nevada. Neues Jahrb. Geol. u. Paläontol. Mh. Pp. 193–194.

ALEKSEEVA, R. E.
  1960. On the genus *Spirigerina* Orbigny. Paleontol. Zh., 4:63–68, pl. 7.
  1962. Devonian Atrypidae of the Kuznetsk and Minusinsk Basins and the east slope of the north Ural. Akad. nauk. SSSR, Siberian Div., Inst., Geol. Geophys. Moscow. Pp. 1–196, pls. 1–12.
  1968. *Sibirispira*, a new genus of the order Atrypida. Akad. nauk. SSSR Doklady, 179, 1:198–201, 1 pl.

AMSDEN, T. W.
  1949. Stratigraphy and Paleontology of the Brownsport Formation (Silurian) of western Tennessee. Peabody Mus. Nat. Hist. Bull. 5. 138 pp., 34 pls.
  1951. Brachiopods of the Henryhouse Formation (Silurian) of Oklahoma. J. Paleontol., 25, 1:69–96, pls. 15–20.
  1958. Stratigraphy and paleontology of the Hunton Group in the Arbuckle Mountain Region; Part II, Haragan articulate brachiopods; Part III, Supplement to the Henryhouse brachiopods. Oklahoma Geol. Surv. Bull. 78. 157 pp., 14 pls.
  1968. Articulate brachiopods of the St. Clair Limestone (Silurian), Arkansas, and the Clarita Formation (Silurian), Oklahoma. Paleontol. Soc. Mem. 1 (J. Paleontol., vol. 42, suppl. to no. 3). vi + 117 pp., 20 pls.

ANDRONOV, S. M.
  1961. On some representatives of the family Pentameridae from the Devonian deposits in the region of the north Urals. Akad. nauk. SSSR, Trudy Geol. Inst., vol. 55. 136 pp., 32 pls.

BARRANDE, JOACHIM
  1847. Über die brachiopoden der silurischen Schichten von Böhmen. Naturwissenschaftliche Abhandl., 1:357–475, pls. 14–22.
  1848. Über die Brachiopoden der silurischen Schichten von Böhmen. Naturwissenschaftliche Abhandl., 2, 2:153–256, pl. 15–23.
  1879. Système Silurien du centre de la Bohême, vol. 5, Brachiopodes. Prague, Paris. 226 pp., 153 pls.

BASSETT, M. G.
  1970. The articulate brachiopods from the Wenlock Series of the Welsh Borderland and South Wales, Part 1, Palaeontol. Soc. Monogr. Pp. 1–26, pls. 1–3.

BERDAN, J. M., and others
  1969. Siluro-Devonian Boundary in North America. Geol. Soc. Amer. 80:2165–2174.

BERRY, W. B. N., and A. J. BOUCOT
  1970. Correlation of the North American Silurian Rocks. Geol. Soc. Amer. Spec. Paper 102. Pp. 1–289.

BERRY, W. B. N., H. JAEGER, and M. A. MURPHY
  1971. The position of *Monograptus uniformis* in stratigraphic sequences in central Nevada. Geol. Soc. Amer. Bull., 82, 7:1969–1972.

BIERNAT, GERTRUDA
  1959. Middle Devonian Orthoidea of the Holy Cross Mountains and their ontogeny. Paleontol. Polonica, no. 10. 78 pp., 12 pls.

BOUCOT, A. J.
  1957. A Devonian brachiopod, *Cyrtinopsis*, redescribed. Senck. leth., 38, ½:37–48, 2 pls.
  1962. Observations regarding some Silurian and Devonian spiriferoid genera. Senck. leth., 43, 5:411–432, pls. 49–52.

BOUCOT, A. J., and T. W. AMSDEN
  1958. New genera of Brachiopods. Oklahoma Geol. Surv. Bull. 78. Pp. 159–170.

BOUCOT, A. J., E. D. GILL, J. G. JOHNSON, A. C. LENZ, and J. A. TALENT
   1966. *Skenidioides* and *Leptaenisca* in the Lower Devonian of Australia (Victoria-Tasmania) and New Zealand, with notes on other Devonian occurrences of *Skenidioides*. Proc. Roy. Soc. Victoria, 79, 2:363–369, pl. 40.
BOUCOT, A. J., and J. G. JOHNSON
   1967. Paleogeography and correlation of Appalachian Province Lower Devonian sedimentary rocks. Tulsa Geol. Soc. Dig., 35:35–87, 2 pls.
   In press. Silurian brachiopods. *In* A. Hallam, ed., Atlas of Palaeobiography. Elsevier, N.Y.
BOUCOT, A. J., J. G. JOHNSON, and J. A. TALENT
   1969. Early Devonian brachiopod zoogeography. Geol. Soc. Amer. Spec. Paper 119. 113 pp., 20 pls.
BOUCOT, A. J., J. G. JOHNSON, and V. G. WALMSLEY
   1965. Revision of the Rhipidomellidae (Brachiopoda) and the affinities of *Mendacella* and *Dalejina*. J. Paleontol. 39, 3:331–340, pls. 45, 46.
BOUCOT, A. J., A. MARTINSSON, R. THORSTEINSSON, O. H. WALLISER, H. B. WHITTINGTON, and E. YOCHELSON
   1960. A late Silurian fauna from the Sutherland River Formation, Devon Island, Canadian Arctic Archipelago. Can. Geol. Surv. Bull. 65. 51 pp., 10 pls.
HAVLÍČEK, VLADIMÍR
   1959a. Spiriferidae v. českém siluru a devonu. Ústřed. Ústavu Geol. Rozpravy, vol. 25. 275 pp., 28 pls.
   1959b. Rhynchonellacea im böhmischen älteren Paläozoikum (Brachiopoda). Ústřed. Ústavu Geol. Věstník, 34:78–82.
   1961. Rhynchonelloidea des böhmischen älteren Paläozoikums (Brachiopoda). Ústřed. Ústavu Geol. Rozpravy, vol. 27. 211 pp., 27 pls.
   1965. Superfamily Orthotetacea (Brachiopoda) in the Bohemian and Moravian Paleozoic. Ústřed. Ústavu Geol. Věstník, 40, 4:291–294.
   1967. Brachiopoda of the suborder Strophomenidina in Czechoslovakia. Ústřed. Ústavu Geol. Rozpravy, vol. 33. 235 pp. 52 pls.
   1971. Non-costate and weakly costate Spiriferidina (Brachiopoda) in the Silurian and Lower Devonian of Bohemia. Shornik geol. Věd (Paleontol.), 14:7–34, 8 pls.
JOHNSON, J. G.
   1965. Lower Devonian stratigraphy and correlation, northern Simpson Park Range, Nevada. Bull. Can. Petrol. Geol., 13, 3:365–381.
   1966. Middle Devonian brachiopods from the Roberts Mountains, central Nevada. Palaeontol., 9, 1:152–181, pls. 23–27.
   1970. Great Basin Lower Devonian Brachiopoda. Geol. Soc. Amer. Mem. 121. xi+421 pp., 74 pls.
JOHNSON, J. G., and A. J. BOUCOT
   1967. *Gracianella*, a new Late Silurian genus of atrypoid brachiopods. J. Paleontol., 41, 4:868–873, pls. 109, 110.
   1968. Llandovery to Givetian brachiopod zonal sequence in the Silurian and Devonian of central Nevada (abst.). Geol. Soc. Amer., Cordilleran Sect., Program 64th Ann. Meeting. P. 69.
   1970. Brachiopods and age of the Tor Limestone of central Nevada. J. Paleontol., 44, 2:265–269, pl. 54.
   In press. Devonian brachiopods. *In* A. Hallam, ed., Atlas of Palaeobiogeography. Elsevier, N.Y.
JOHNSON, J. G., A. J. BOUCOT, and M. A. MURPHY
   1967. Lower Devonian faunal succession in central Nevada. Alberta Soc. Petrol. Geol., Internat. Symp. on the Devonian System, Calgary, 1967, 2:679–691 (publ. in 1968).
JOHNSON, J. G., and J. A. TALENT
   1967a. *Muriferella*, a new genus of Lower Devonian septate dalmanellid. Proc. Roy. Soc. Victoria, 80, 1:43–50, pls. 9, 10.
   1967b. Cortezorthinae, a new subfamily of Siluro-Devonian dalmanellid brachiopods. Palaeontol., 10, 1:142–170, pls. 19–22.

KEMEŽYS, K. J.
  1968. Arrangements of costellae, setae and vascula in enteletacean brachiopods. J. Paleontol., 42, 1:88–93.

KHODALEVICH, A. N.
  1939. Upper Silurian Brachiopoda of eastern slope of the Ural. Geol. Serv. USSR. Trans. Ural Geol. Serv. 135 pp., 28 pls.
  1951. Lower Devonian and Eifelian brachiopods of the Ivdel and Serov districts of the Sverdlovsk region. Trudy, Sverdlovsk Mining Inst., vol. 18. 169 pp., 30 pls.

KOZLOWSKI, ROMAN
  1929. Les Brachiopodes Gothlandiens de la Podolie Polonaise. Palaeontol. Polonica, vol. 1. 254 pp., 12 pls.

KULKOV, N. P.
  1963. Brachiopodi Solovikhinskikh Sloev Nizhnego Devona gornogo Altaia. Akad. nauk. SSSR. 132 pp., 9 pls.
  1966. The genus *Septatrypa* Kozlowski, 1929. Doklady Akad. nauk SSSR, 167, 1:191–193.

LAZAREV, S. S.
  1970. Morphology and systematics of the brachiopod superfamily Enteletacea. Paleontol. Inst. SSSR, Paleontol. and Strat. no. 128. Moscow. Pp. 1–31.

LENZ, A. C.
  1967. Upper Silurian and Lower Devonian biostratigraphy, Royal Creek, Yukon Territory, Canada. Alberta Soc. Petrol. Geol., Internat. Symposium on the Devonian System, Calgary, 1967, 2:587–599 (publ. in 1968).
  1970. Late Silurian brachiopods of Prongs Creek, northern Yukon. J. Paleontol., 44, 3:480–500, pls. 83–87.

McLAREN, D. J., A. W. NORRIS, and D. C. McGREGOR
  1962. Illustrations of Canadian fossils: Devonian of Western Canada. Can. Geol. Surv. Paper 62–64. 34 pp., 16 pls.

MERRIAM, C. W.
  1940. Devonian stratigraphy and paleontology of the Roberts Mountains region, Nevada. Geol. Soc. Amer. Spec. Paper 25. 114 pp., 16 pls.
  1963. Paleozoic rocks of Antelope Valley, Eureka and Nye Counties, Nevada. U.S. Geol. Surv., Prof. Paper 423. 67 pp., 2 pls.

NIKIFOROVA, O. I.
  1937. Upper Silurian Brachiopoda of the central Asiatic part of the USSR. Centr. Geol. Prospect. Inst., Monogr. Paleontol. USSR, vol. 35. 93 pp., 14 pls.
  1954. Stratigraphy and brachiopods of the Silurian deposits of Podolia. Trudy VSEGEI. 218 pp., pt. 1, 19 pls.; pt. 2, 18 pls.
  1970. Brachiopods of the Greben Horizon of Vaigatsch (latest Silurian). *In* S. V. Cherkesova, ed., Stratigraphy and faunas of the Silurian deposits of Vaigatsch. Sci. Inst. Geol. Arctic, Min. Geol. SSSR. Pp. 97–149, 7 pls.

RUBEL, M., and T. MODZALEVSKAYA
  1967. New Silurian brachiopods of the family Athyrididae. Izvestia Akad. Nauk, Eston SSR, 16 (Khimiia-Geologiia 1967), 3:238–241.

RUKAVISHNIKOVA, T. B.
  1961. Brachiopods of the upper Silurian of the northern Pribalkhasch. *In* Materials for the geology and guide fossils of Kazakhstan, pt. 1 (26), Stratigraphy and paleontol. Min. Geol. Nedr. Kazakh SSR. Pp. 38–63, pls. 1–6.

RZHONSNITSKAYA, M. A.
  1960. Order Atrypida. *In* T. G. Sarycheva, Osnovi Paleontologii, Mshanki, Brachiopodi. Izdat. Akad. Nauk SSSR. Pp. 257–264, pls. 53–56.

SAVAGE, N. M.
  1970. New atrypid brachiopods from the Lower Devonian of New South Wales. J. Paleontol., 44, 4:655–668, pls. 101–103.

SIEHL, AGEMAR
  1962. Der Greifensteiner Kalk (Eiflium, Rhenisches Schiefergebirge) und seine Brachio-
        podenfauna. 1. Geologie, Atrypacea und Rostrospiracea. Palaeontographica, 119, A:
        173–221, pls. 23–40.
STAUFFER, C. R.
  1930. The Devonian of California. Calif. Univ. Publ. in Geol. Sci., 19, 4:81–118, pls. 10–14.
TSCHERNYSCHEW, TH.
  1893. Die fauna des unteren Devon am ostabhange des Ural. Mém. Comité Geol. St.-Peters-
        bourg. 221 pp., 14 pls.
WALMSLEY, V. G.
  1965. *Isorthis* and *Salopina* (Brachiopoda) in the Ludlovian of the Welsh Borderland.
        Palaeontol., 8, 3:454–477, pls. 61–65.
WALMSLEY, V. G., and A. J. BOUCOT
  1971. The Resserellinae: A new subfamily of Late Ordovician to Early Devonian dalmanellid
        brachiopods. Palaeontol., 14, 3:487–531, pls. 91–102.
WALMSLEY, V. G., A. J. BOUCOT, and C. W. HARPER
  1969. Silurian and Lower Devonian salopinid brachiopods. J. Paleontol., 43, 2:492–516, pls.
        71–80.
WILLIAMS, ALWYN, ed.
  1965. Treatise on Invertebrate Paleontology, Part H, Brachiopoda. Geol. Soc. Amer. and
        Univ. of Kansas Press. 927 pp.
WINTERER, E. L., and M. A. MURPHY
  1960. Silurian reef complex and associated facies, central Nevada. J. Geol., 68, 2:117–139,
        7 pls.
WRIGHT, A. D.
  1968. The brachiopod *Dicoelosia biloba* (Linnaeus) and related species. Arkiv för Zoologi,
        ser. 2, 20, 14:261–319, 7 pls.

PLATES

PLATE 1

Figs. 1–8. *Ptychopleurella* sp. E
  Pridolian fauna, UCR 5428.
    1, 2, Exterior and interior of pedicle valve × 3, USNM 171271.
    3, 4, Interior and exterior of brachial valve × 3, USNM 171272.
    5, 6, Interior and exterior of brachial valve × 3, USNM 171273.
    7, 8, Interior and exterior of brachial valve × 4, USNM 171274.

Figs. 9–13. *"Dolerorthis"* sp.
  E fauna, Pridolian, fig. 9, UCR 5430; Pridolian fauna, figs. 10–13, UCR
5428.
    9, Exterior of pedicle valve × 2, USNM 171275.
    10, 11, Exterior and interior of pedicle valve × 3, USNM 171276.
    12, Interior of brachial valve × 2, USNM 171277.
    13, Exterior of brachial valve × 2, USNM 171278.

Figs. 14–21. *Salopina* sp. E
  E fauna, Pridolian, UCR 5430.
    14, 15, Interior and exterior of brachial valve × 3, USNM 171279.
    16, 17, Exterior and interior of brachial valve × 3, USNM 171280.
    18, Interior of brachial valve × 3, USNM 171281.
    19, Exterior of pedicle valve × 3, USNM 171282.
    20, 21, Exterior and interior of pedicle valve × 3, USNM 171591.

Figs. 22–26. *Leptaena* sp. E
  E fauna, Pridolian, UCR 5430.
    22–24, Side, interior, and exterior of pedicle valve × 2.5, USNM 171283.
    25, 26, Exterior and interior of brachial valve × 2.5, USNM 171284.

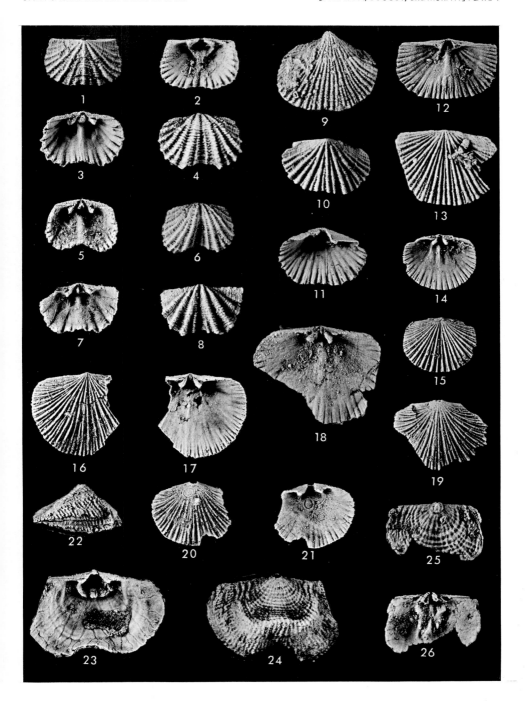

PLATE 2

Figs. 1–18. *Reticulatrypa neutra* n. sp.
   Pridolian fauna, all specimens from UCR 5428 except figs. 13–15 from
UCR 5437.
   1, Exterior of pedicle valve × 3, USNM 171285.
   2, 3, Exterior and interior of brachial valve × 3, USNM 171286.
   4, Exterior of brachial valve × 3, USNM 171287.
   5, Exterior of pedicle valve × 3, USNM 171288.
   6, 7, Exterior and interior of brachial valve × 3, USNM 171289.
   8–12, Side, anterior, posterior, dorsal, and ventral views × 3, USNM
      171290.
   13, Exterior of pedicle valve × 2, USNM 171291.
   14, 15, Exterior and interior of brachial valve × 2, USNM 171292.
   16, Exterior of pedicle valve × 3, USNM 171293.
   17, 18, Exterior and interior of brachial valve × 3, USNM 171294.

Figs. 19–24. *Reticulatrypa* aff. *granulifera* (Barrande)
   Pridolian fauna, figs. 19–21, UCR 5436; figs. 22–24, UCR 5434.
   19–21, Ventral, dorsal, and posterior views × 4, USNM 171295.
   22, 23, Exterior and interior of pedicle valve × 4, USNM 171296.
   24, Interior of brachial valve × 4, USNM 171297.

PLATE 3

Figs. 1–20. *Gracianella cryptumbra* n. sp.
   E fauna, Pridolian UCR 5428.
   1, 2, Posterior and ventral views of pedicle valve × 5, USNM 171298.
   3, Exterior of pedicle valve × 6, USNM 171299.
   4, Exterior of brachial valve × 5, USNM 171300.
   5, Interior of pedicle valve × 6, USNM 171301.
   6, Interior of pedicle valve × 5, USNM 171302.
   7, Interior of brachial valve × 6, USNM 171303.
   8, Exterior of pedicle valve × 5, USNM 171304.
   9–11, Ventral, posterior, and side views × 5, USNM 171305.
   12–14, Ventral, anterior, and dorsal views × 6, USNM 171306.
   15, 16, Exterior and interior of brachial valve × 5, USNM 171307.
   17, 18, Dorsal and ventral views of exterior × 5, USNM 171308.
   19, 20, Ventral and dorsal views of exterior × 5, USNM 171309.

Figs. 21–26. *Delthyris* sp.
   Pridolian fauna, UCR 5428.
   21, Exterior of pedicle valve × 4, USNM 171310.
   22, Exterior of brachial valve × 3, USNM 171311.
   23, 24, Exterior and interior of pedicle valve × 3, USNM 171312.
   25, 26, Interior and exterior of brachial valve × 4, USNM 171313.

PLATE 4

Figs. 1–20. *Morinorhynchus punctorostra* n. sp.

E fauna, Pridolian, UCR 5430.

1–4, Interior, exterior, posterior, and side views of pedicle valve × 2, USNM 171314.

5, 6, Side view and anterior view of interior of pedicle valve × 2, USNM 171315.

7, Posterior view of pedicle valve × 2, USNM 171316.

8, 9, Interior and exterior views of pedicle valve × 3, USNM 171317.

10, 11, Anterior and posterior views of fragment of pedicle valve × 2, USNM 171318.

12, Exterior of pedicle valve × 3. Note notched ventral beak, USNM 171319.

13, Posterior view of pedicle valve × 2, USNM 171320.

14, Posterior view of pedicle valve × 2, USNM 171321.

15, 19, Interior and posterior views of pedicle valve × 2, USNM 171322.

16, 17, Exterior and interior of pedicle valve × 2, USNM 171323. Note punctured ventral beak in figure 16.

18, Posterior of pedicle valve × 6, USNM 171324. Note multipunctured shell.

20, Interior of pedicle valve × 5, USNM 171325. Note preservation of the rough attachment surface of the adductor muscles.

PLATE 5

Figs. 1–17. *Morinorhynchus punctorostra* n. sp.
  E fauna, Pridolian, UCR 5430.
  1, 2, Posterior views of brachial valve × 12 and × 4, USNM 171326.
  3, Posterior view of brachial valve × 4, USNM 171327.
  4–8, Exterior, interior, anterior, posterior, and side views × 1.5, USNM
    171328.
  9, Exterior of brachial valve × 3, USNM 171329.
  10, Exterior of brachial valve × 3, USNM 171330.
  11, Exterior of brachial valve × 2, USNM 171331.
  12–16, Side, anterior, posterior, interior, and exterior of brachial valve × 2,
    USNM 171332.
  17, Exterior of brachial valve × 3, USNM 171333.

PLATE 6

Figs. 1–18. *Mesodouvillina costatuloides* n. sp.

    E fauna, Pridolian, UCR 5430.

    1, Interior of brachial valve × 3, USNM 171334.

    2, Interior of brachial valve × 3, USNM 171335.

    3, Exterior of pedicle valve × 2, USNM 171336.

    4–8, Side, dorsal, ventral, posterior, and anterior views × 2, USNM 171337.

    9, 10, Side and interior views of pedicle valve × 2, USNM 171338.

    11, 12, Exterior and side views × 3, USNM 171339.

    13, Interior of pedicle valve × 3, USNM 171340.

    14–16, Interior and exterior of pedicle valve × 3, and margin of ventral exterior × 10, USNM 171341.

    17, 18, Interior and exterior of pedicle valve × 3, USNM 171342.

PLATE 7

Figs. 1–18. *Gracianella reflexa* n. sp.
  E fauna, Pridolian, UCR 5430.
  1–3, Exterior, interior, and anterior views of pedicle valve × 5, USNM
    171343.
  4, Anterior view of brachial valve × 5, USNM 171344.
  5, 6, Exterior and interior of pedicle valve × 5, USNM 171345.
  7, Anterior view of articulated valves with brachial valve above, × 5, USNM
    171346.
  8, Interior of brachial valve × 5, USNM 171347.
  9–13, Ventral, posterior, dorsal, anterior, and side views × 4, USNM 171348.
  14, Interior of brachial valve × 5, USNM 171349.
  15, 16, Interior and exterior of brachial valve × 5, USNM 171350.
  17, 18, Dorsal and ventral views of exterior × 4, USNM 171351.

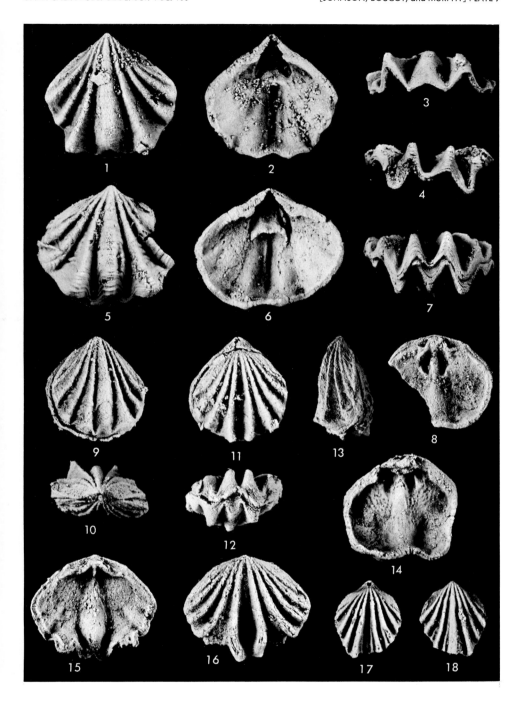

PLATE 8

Figs. 1–10. *Gracianella reflexa* n. sp.

    E fauna, Pridolian, UCR 5430.

    1, Anterior of brachial valve × 7, USNM 171352.

    2, Interior of brachial valve × 6, USNM 171353.

    3–7, Ventral, posterior, dorsal, anterior, and side views × 4, USNM 171354.

    8, Anterior view of articulated valves in which marginal plication has become obsolete × 5, USNM 171355.

    9, Interior of brachial valve × 6, USNM 171356. Note elevated marginal rim developed as in the specimen in figure 8.

    10, Posterior view of exterior × 5, USNM 171357.

Figs. 11–15. *Atrypella?* sp.

    E fauna, Pridolian, UCR 5430.

    Ventral, dorsal, posterior, anterior, and side views × 2, USNM 171358.

Figs. 16–25. *Protathyris* sp. E

    E fauna, Pridolian, UCR 5430.

    16–20, Exterior, interior, side, anterior, and posterior views of brachial valve × 5, USNM 171359.

    21, Interior of pedicle valve × 5, USNM 171360.

    22, 23, Interior and posterior of pedicle valve × 4, USNM 171361.

    24, 25, Exterior and interior of brachial valve × 5, USNM 171362.

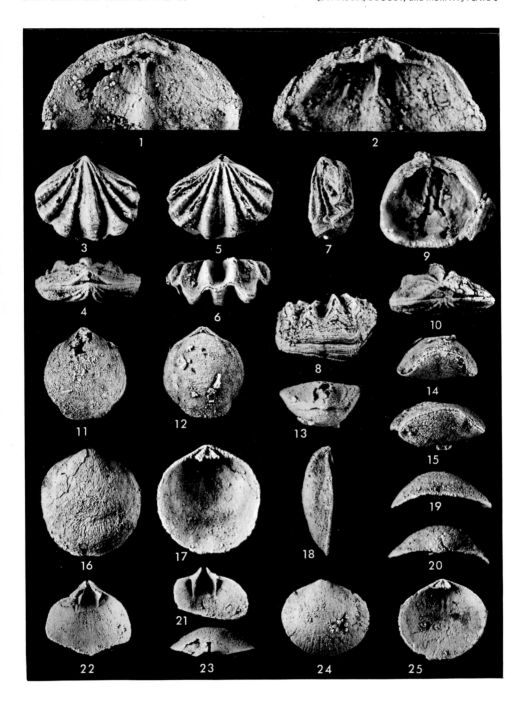

PLATE 9

Figs. 1–19. *Tenellodermis matrix* n. sp.
   E fauna, Pridolian, UCR 5430.
   1–3, Interior × 5, side × 3, and exterior × 3 of brachial valve, USNM 171363.
   4–6, Posterior, exterior, and interior of brachial valve × 3, USNM 171364.
   7, 8, Interior and exterior of brachial valve × 3, USNM 171365.
   9–11, Exterior, posterior, and anterior views of pedicle valve × 3, USNM
      171366.
   12, Interior of pedicle valve × 3, USNM 171367.
   13, Exterior of brachial valve × 3, USNM 171368.
   14–16, Anterior and exterior × 3 and anterior × 5 of pedicle valve, USNM
      171369.
   17–19, Side, posterior, and exterior views of pedicle valve × 3, USNM
      171370.

PLATE 10

Figs. 1–13. *Skenidioides robertsensis* n. sp.

  F fauna, Gedinnian, UCR 5462.

  1–5, Ventral, dorsal, side, posterior, and anterior views × 4, USNM 171371.

  6, Exterior of pedicle valve × 5, USNM 171372.

  7–9, Dorsal, side, and posterior views of exterior × 5, USNM 171373.

  10, 11, Posterior and interior views of pedicle valve × 4, USNM 171374.

  12, Interior of brachial valve × 5, USNM 171375.

  13, Interior of brachial valve × 5, USNM 171376.

Figs. 14–18. *Ptychopleurella* sp. F

  F fauna, Gedinnian, figs. 14–16, UCR 5443; figs. 17, 18, UCR 5442.

  14, 15, Exterior and interior of brachial valve × 3, USNM 171377.

  16, Exterior of pedicle valve × 3, USNM 171378.

  17, 18, Interior and exterior of pedicle valve × 2, USNM 171379.

Figs. 19–34. *Schizophoria paraprima* n. sp.

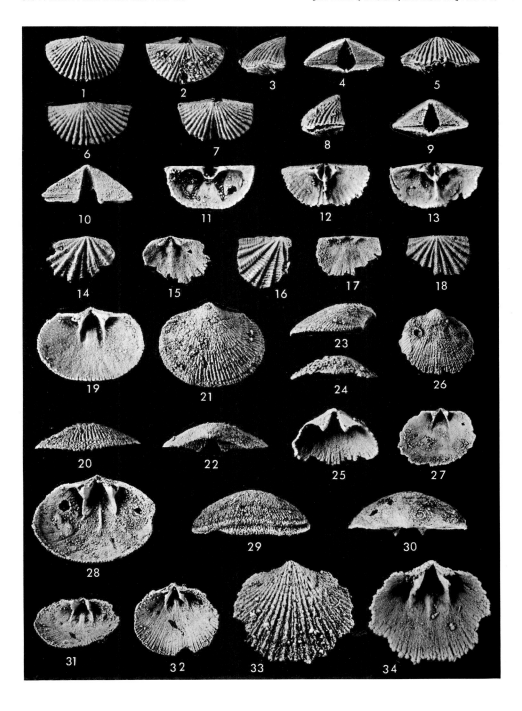

PLATE 11

Figs. 1–11. *Schizophoria paraprima* n. sp.

F fauna, Gedinnian, fig. 1, UCR 5449; figs. 2–6, UCR 5462; figs. 7–11, UCR 5461.

1, Interior of pedicle valve × 3, USNM 171386.

2, Anterior view of interior of brachial valve × 7, USNM 171387.

3, 4, Exterior and posterior of brachial valve × 7, USNM 171388.

5, 6, Two views of interior of brachial valve × 10, USNM 171389.

7–9, Posterior, exterior, and interior of pedicle valve × 2, USNM 171390.

10, 11, Two views of interior of brachial valve × 4, USNM 171391.

Figs. 12–14. *Schizophoria* cf. *fragilis* Kozlowski

F fauna, Gedinnian, fig. 12, UCR 5463; figs. 13, 14, UCR 5460.

12, Interior of fragment of pedicle valve × 5, USNM 171392.

13, Interior of fragment of pedicle valve × 4, USNM 171393.

14, Interior of fragment of brachial valve × 4, USNM 171394.

Figs. 15–23. *Salopina submurifer* n. sp.

F fauna, Gedinnian, UCR 5462.

15, 16, Dorsal and ventral views of exterior × 5, USNM 171395.

17, Anterior view of interior of brachial valve × 7, USNM 171396.

18, Anterior view of brachial valve × 5, USNM 171397.

19–23, Ventral, dorsal, side, posterior, and anterior views × 5, USNM 171398.

PLATE 12

Figs. 1–19. *Salopina submurifer* n. sp.
  F fauna, Gedinnian, figs. 1, 2, UCR 5461; figs. 3–19, UCR 5462.
  1, Interior of brachial valve × 5, USNM 171399.
  2, Anterior view of interior of brachial valve × 5, USNM 171400.
  3, Interior of brachial valve × 4, USNM 171401.
  4, Interior of brachial valve × 5, USNM 171402.
  5, Interior of brachial valve × 4, USNM 171403.
  6, Interior of brachial valve × 4, USNM 171404.
  7, Interior of brachial valve × 5, USNM 171405.
  8, Interior of brachial valve × 5, USNM 171406.
  9, Interior of brachial valve × 7, USNM 171407.
  10, 11, Exterior and interior of brachial valve × 7, USNM 171408.
  12, 13, Interior and exterior of brachial valve × 5, USNM 171409.
  14, 15, Interior and exterior of brachial valve × 5, USNM 171410.
  16, Anterior view of interior of brachial valve × 5, USNM 171411.
  17, Interior of pedicle valve × 5, USNM 171412.
  18, Exterior of pedicle valve × 5, USNM 171413.
  19, Interior of pedicle valve × 5, USNM 171414.

Figs. 20–24. *Dalejina subfrequens* n. sp.
  F fauna, Gedinnian, UCR 5458.
  Dorsal, ventral, side, anterior, and posterior views × 2.5, USNM 171415.

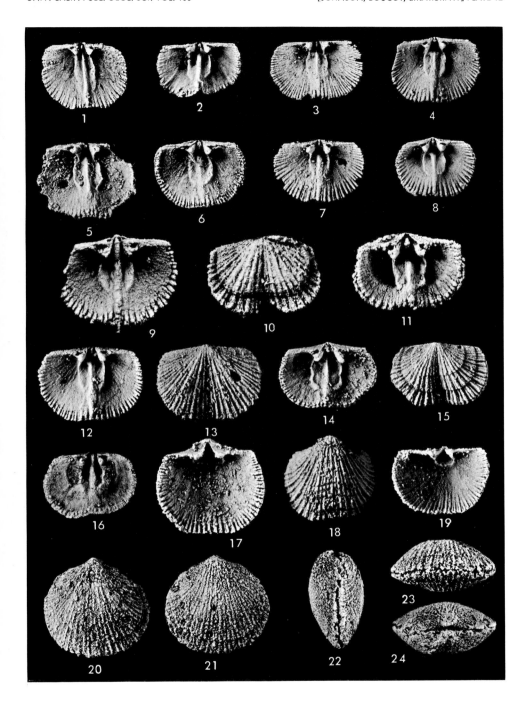

PLATE 13

Figs. 1–24. *Dalejina subfrequens* n. sp.

F fauna, Gedinnian, figs. 1–10, 14, 20–24, UCR 5458; figs. 11–13, 15–19, UCR 5453.

1, 2, Interior and exterior of pedicle valve × 3, USNM 171416.

3, 4, Interior and exterior of pedicle valve × 3, USNM 171417.

5, 6, Two views of interior of brachial valve × 1.5, USNM 171418.

7, 8, Exterior and interior of brachial valve × 4, USNM 171419.

9, 10, Interior and exterior of pedicle valve × 3, USNM 171420.

11, 12, Exterior and interior of brachial valve × 3, USNM 171421.

13, Exterior of pedicle valve × 3, USNM 171422.

14, Interior of pedicle valve × 1.5, USNM 171423.

15–19, Ventral, dorsal, side, anterior, and posterior views × 3, USNM 171424

20–24, Exterior, interior, posterior, anterior, and side views of brachial valve × 2, USNM 171425.

Figs. 1–21. *Resserella elegantuloides* (Kozlowski)

F fauna, Gedinnian, UCR 5458.

1–5, Exterior, interior, side, posterior, and anterior views × 2.5, USNM 171426.

6–8, Exterior, interior, and anterior views of pedicle valve × 4, USNM 171427.

9, 10, Interior and exterior of brachial valve × 4, USNM 171428.

11, 12, Exterior and interior of pedicle valve × 3, USNM 171429.

13–16, Posterior, exterior, interior, and anterior views of brachial valve × 3, USNM 171430.

17–19, Anterior, exterior, and interior of brachial valve × 3, USNM 171431.

20, 21, Posterior and side views × 3, USNM 171432.

PLATE 15

Figs. 1–23. *Tyersella jubar* n. sp.
F fauna, Gedinnian, figs. 1–19, UCR 5460; figs. 20–23, USNM 10795.
1–5, Ventral, dorsal, anterior, posterior, and side views × 2, USNM 171433.
6–10, Ventral, dorsal, posterior, anterior, and side views × 3, USNM 171434.
11–15, Ventral, anterior, side, posterior, and dorsal views × 1.5, USNM 171435.
16, Interior of pedicle valve × 1.5, USNM 171436.
17, Interior of pedicle valve × 3, USNM 171437.
18, Anterior view of interior of pedicle valve × 5, USNM 171438.
19, Interior of pedicle valve × 1.5, USNM 171439.
20–23, Posterior, side, exterior, and interior of pedicle valve × 2, USNM 171440.

PLATE 16

Figs. 1–15. *Tyersella jubar* n. sp.

F fauna, Gedinnian, figs. 1–7, 10, 11, USNM 10795; figs. 8, 9, 12–15, UCR 5460.

1–4, Interior, exterior, posterior, and side views of brachial valve × 2, USNM 171441.

5, Interior of brachial valve × 2, USNM 171442.

6, Interior of brachial valve × 1.5, USNM 171443.

7, Interior of brachial valve × 3, USNM 171444.

8, Interior of brachial valve × 10, USNM 171445.

9, Interior of brachial valve × 10, USNM 171446.

10, Interior of brachial valve × 2, USNM 171447.

11, Interior of pedicle valve × 2, USNM 171448.

12, 13, Two views of interior of brachial valve × 4, USNM 171449.

14, 15, Two views of interior of brachial valve × 4, USNM 171450.

Figs. 16–24. *Dicaelosia nitida* n. sp.

F fauna, Gedinnian, figs. 16–19, UCR 5462; figs. 20–24, UCR 5461.

16, 17, 19, Ventral, dorsal, and side views × 4, USNM 171451.

18, Interior of pedicle valve × 3, USNM 171452.

20–24, Ventral, dorsal, side, posterior, and anterior views × 5, USNM 171453.

F fauna, Gedinnian, UCR 5462, except figs. 26, 27, UCR 5461.

19–23, Interior, anterior, exterior, posterior, and side views × 4, USNM 171380.

24, 25, Anterior and posterior views of pedicle valve × 5, USNM 171381.

26, 27, Exterior and interior of pedicle valve × 2, USNM 171382.

28–30, Interior, anterior, and posterior views of brachial valve × 3, USNM 171383.

31, 32, Anterior and interior views of brachial valve × 3, USNM 171384.

33, 34, Exterior and interior of brachial valve × 6, USNM 171385.

PLATE 17

Figs. 1–7. *Dicaelosia nitida* n. sp.

    F fauna, Gedinnian, figs. 1–3, UCR 5461; figs. 4–7, UCR 5462.

    1, Interior of pedicle valve × 5, USNM 171454.

    2, Interior of brachial valve × 5, USNM 171455.

    3, Interior of brachial valve × 5, USNM 171456.

    4, 5, Exterior and interior of brachial valve × 5, USNM 171457.

    6, 7, Interior and exterior of pedicle valve × 5, USNM 171458.

Figs. 8–22. *Anastrophia magnifica* Kozlowski

    F fauna, Gedinnian, figs. 8, 9, UCR 5459; figs. 10–19, UCR 5458; figs. 20–22, UCR 5469.

    8, 9, Exterior and interior of pedicle valve × 2.5, USNM 171459.

    10, Interior of pedicle valve × 2.5, USNM 171460.

    11–14, Exterior, side, posterior, and interior views of brachial valve × 2, USNM 171461.

    15–19, Dorsal, ventral, posterior, side, and anterior views × 2.5, USNM 171462.

    20, 21, Interior and exterior of pedicle valve × 2, USNM 171463.

    22, Interior of posterior fragment of brachial valve × 4, USNM 171464.

PLATE 18

Figs. 1–18. *Gypidula pelagica lux* n. subsp.

F fauna, Gedinnian all UCR 5459 except figs. 15, 16, which are UCR 5460.

1–5, Ventral, dorsal, posterior, side, and anterior views × 1.25, USNM 157016.

6–10, Anterior, ventral, posterior, dorsal, and side views × 1.5, USNM 171465.

11–14, Ventral, dorsal, side, and posterior views × 1.25, USNM 171466.

15, 16, Anterior and interior views of brachial valve × 1.25, USNM 171467.

17, 18, Anterior and ventral views × 1.25, USNM 171468.

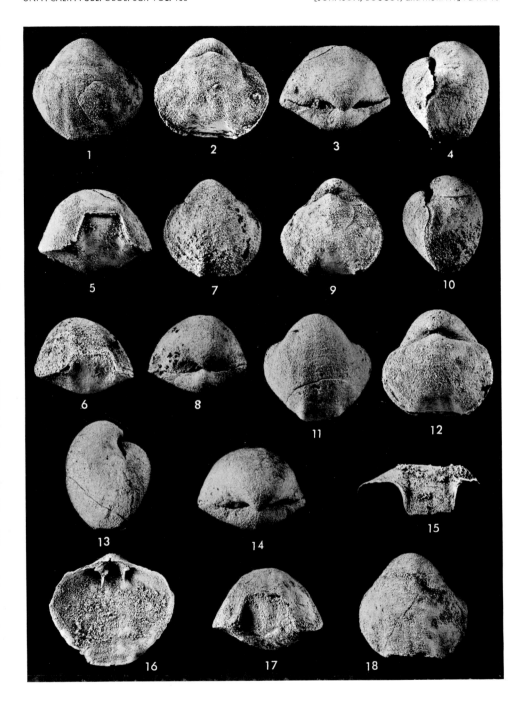

PLATE 19

Figs. 1–12. *Gypidula pelagica lux* n. subsp.
    F fauna, Gedinnian, UCR 5459.
    1–5, Ventral, dorsal, side, posterior, and anterior views × 2, USNM 171469.
    6–10, Ventral, dorsal, side, anterior, and posterior views × 2.5, USNM 171470.
    11, Interior of pedicle valve × 1.5, USNM 171471.
    12, Interior of pedicle valve × 1.5, USNM 171472.

Figs. 13–18. *Gypidula?* sp.
    F fauna, Gedinnian, UCR 5461.
    13, Interior of brachial valve × 2, USNM 171473.
    14, Interior of brachial valve × 3, USNM 171474.
    15, Interior of brachial valve × 5, USNM 171475.
    16, Interior of brachial valve × 4, USNM 171476.
    17, Oblique view of interior of brachial valve × 4, USNM 171477.
    18, Interior of brachial valve × 4, USNM 171478.

PLATE 20

Figs. 1–7. *Gypidula* sp. F

    F fauna, Gedinnian, figs. 1, 2, UCR 5461; figs. 3, 4, UCR 5466; figs. 5–7, UCR 5469.

    1, 2, Interior and exterior of brachial valve × 2, USNM 171479.

    3, 4, Interior and exterior of brachial valve × 2, USNM 171480.

    5–7, Interior, exterior, and posterior views of pedicle valve × 2, USNM 171481.

Figs. 8–11. *Leptaenisca* sp.

    F fauna, Gedinnian, fig. 8, UCR 5462; figs. 9, 10, UCR 5453; fig. 11, UCR 5469.

    8, Interior of brachial valve × 3, USNM 171482.

    9, 10, Interior and posterior views of brachial valve × 3, USNM 171483.

    11, Interior of pedicle valve × 2, USNM 171484.

Figs. 12–14. *Leptaena* sp. F

    F fauna, Gedinnian, UCR 5458.

    12, 13, Dorsal and ventral views × 1.25, USNM 171485.

    14, Exterior of pedicle valve × 1.5, USNM 171486.

Figs. 15, 16. *Lepidoleptaena* sp.

    F fauna, Gedinnian, fig. 15, UCR 5457; fig. 16, UCR 5458.

    15, Interior of pedicle valve assembled from two separate fragments × 1, USNM 171487.

    16, Interior of brachial valve × 1, USNM 171488.

Figs. 17–23. *Aesopomum varistriatus* Johnson

    F fauna, Gedinnian, figs. 17–19, UCR 5462; figs. 20–23, UCR 5465.

    17, 18, Posterior and interior views of brachial valve × 3, USNM 171489.

    19, Anterior view of posterior fragment of interior of pedicle valve × 3, USNM 171490.

    20–23, Exterior, interior, posterior, and side views of pedicle valve × 4, USNM 171491.

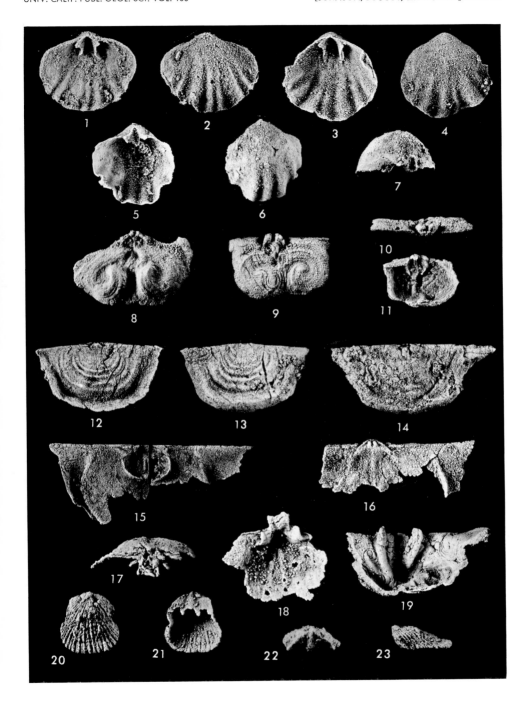

PLATE 21

Figs. 1–8. *Iridistrophia* cf. *umbella* (Barrande)
   F fauna, Gedinnian, UCR 5458.
   1, 2, Two views of interior of pedicle valve × 1.5, USNM 171492.
   3, 4, Exterior and interior of fragment of pedicle valve × 1.25, USNM
      171493.
   5–8, Exterior, interior, side, and posterior views of brachial valve × 1.5,
      USNM 171494.

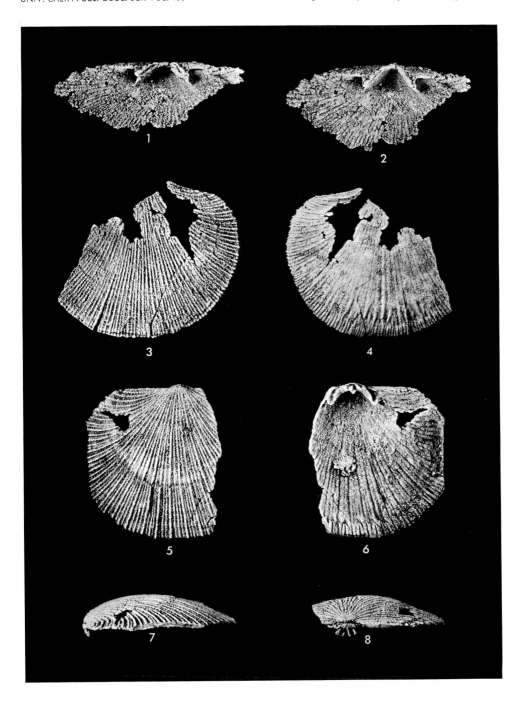

PLATE 22

Figs. 1–19. *Ancillotoechia gutta* n. sp.

  F fauna, Gedinnian, UCR 5459.

  1–5, Ventral, dorsal, posterior, anterior, and side views × 3, USNM 171495.

  6–9, Side, anterior, ventral, and dorsal views × 2, USNM 171496.

  10–14, Dorsal, ventral, side, anterior, and posterior views × 3, USNM 171497.

  15, 16, Ventral and dorsal views × 3, USNM 171498.

  17, Interior of brachial valve × 7, USNM 171499.

  18, Interior of brachial valve × 7, USNM 171500.

  19, Interior of brachial valve × 7, USNM 171501.

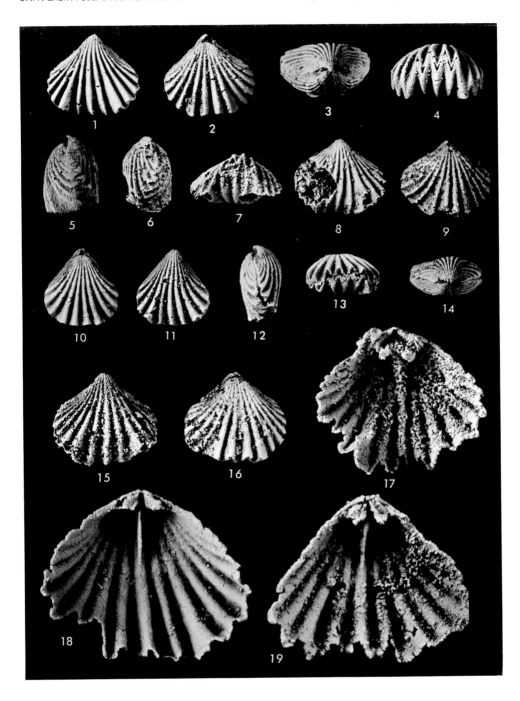

Figs. 1–11. *Sphaerirhynchia gibbosa* (Nikiforova)

    F fauna, Gedinnian, UCR 5461.

    1–5, Ventral, anterior, dorsal, posterior, and side views × 2, USNM 171502.

    6, Interior of pedicle valve × 2, USNM 171503.

    7, Interior of brachial valve × 5, USNM 171504. Note inner hinge plates.

    8–10, Side, anterior, and posterior views × 2, USNM 171505.

    11, Interior of posterior portions of articulated valves × 5, USNM 171506.

Figs. 12–16, *Hebetoechia?* cf. *ornatrix* Havilíček

    F fauna, Gedinnian, UCR 5458.

    Exterior, interior, posterior, anterior, and side views of pedicle valve × 1.5, USNM 171507.

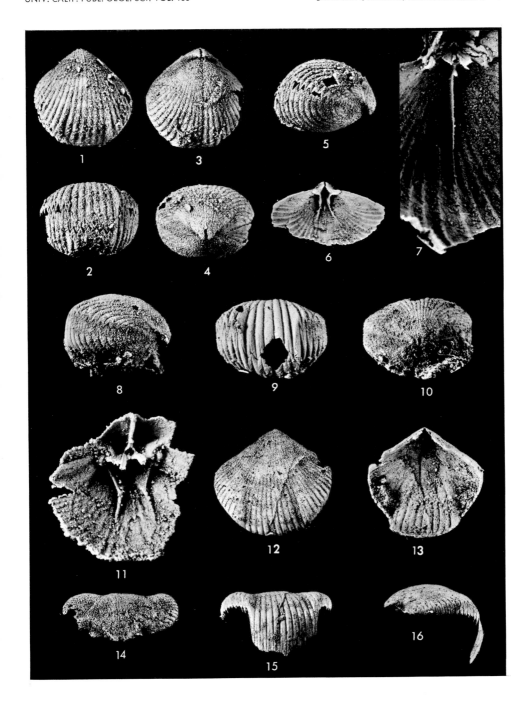

PLATE 24

Figs. 1–13. *Eoglossinotoechia* cf. *cacuminata* Havlíček
    F fauna, Gedinnian, figs. 1–8, UCR 5462; figs. 9–13, UCR 6306.
    1–5, Interior, exterior, posterior, anterior, and side views of pedicle valve ×
        3, USNM 171508.
    6, 7, Anterior and interior views of brachial valve × 5, USNM 171509.
    8, Interior of pedicle valve × 2, USNM 171510.
    9, Anterior view × 4, USNM 171511.
    10, 11, Interior and exterior of pedicle valve × 4, USNM 171512.
    12, 13, Exterior and interior of brachial valve × 4, USNM 171513.

Figs. 14–27. *Atrypa nieczlawiensis* Kozlowski
    F fauna, Gedinnian, figs. 14–20, UCR 5458; figs. 21–27, UCR 5462.
    14, 15, Anterior and dorsal views × 1.5, USNM 171514.
    16, 20, Ventral, posterior, dorsal, anterior, and side views × 1.25, USNM
        171515.
    21, Dorsal views × 2, USNM 171516.
    22, Ventral view × 1.5, USNM 171517.
    23, Ventral view × 1.5, USNM 171518.
    24, Ventral view × 2, USNM 171519.
    25–27, Posterior, ventral, and dorsal views × 4, USNM 171520.

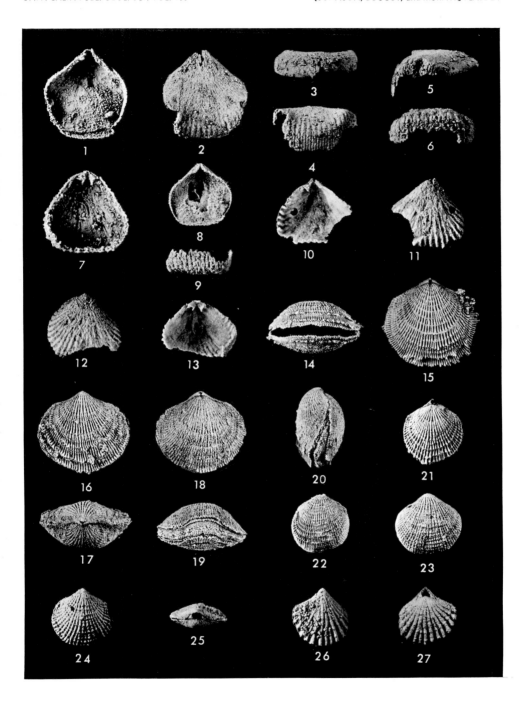

PLATE 25

Figs. 1–7. *Spirigerina marginaliformis* Alekseeva
  F fauna, Gedinnian, UCR 5445.
  1, 2, Interior and exterior of pedicle valve × 2, USNM 171521.
  3, Exterior of pedicle valve × 2, USNM 171522.
  4, Exterior of pedicle valve × 2, USNM 171523.
  5–7, Dorsal, ventral, and side views × 2, USNM 171524.

Figs. 8–11. *Sibirispira?* sp.
  F fauna, Gedinnian, UCR 5444.
  8, 9, Ventral and dorsal views × 4, USNM 171525.
  10, Ventral view × 4, USNM 171526.
  11, Dorsal view × 4, USNM 171527.

Figs. 12–25. *Atrypina prosimpsoni* n. sp.
  F fauna, Gedinnian, figs. 12, 13, UCR 5458; fig. 14, UCR 5461; fig. 15, UCR 5455; figs. 16–20, UCR 5460; figs. 21, 22, UCR 5461; figs. 23–25, UCR 5465.
  12, 13, Exterior and interior of brachial valve × 4, USNM 171528.
  14, Exterior of brachial valve × 3, USNM 171529.
  15, Interior of pedicle valve × 3, USNM 171530.
  16–20, Exterior, interior, anterior, posterior, and side views of pedicle valve × 4, USNM 171531.
  21, 22, Dorsal and posterior views × 5, USNM 171532.
  23–25, Posterior, ventral, and dorsal views × 5, USNM 171533.

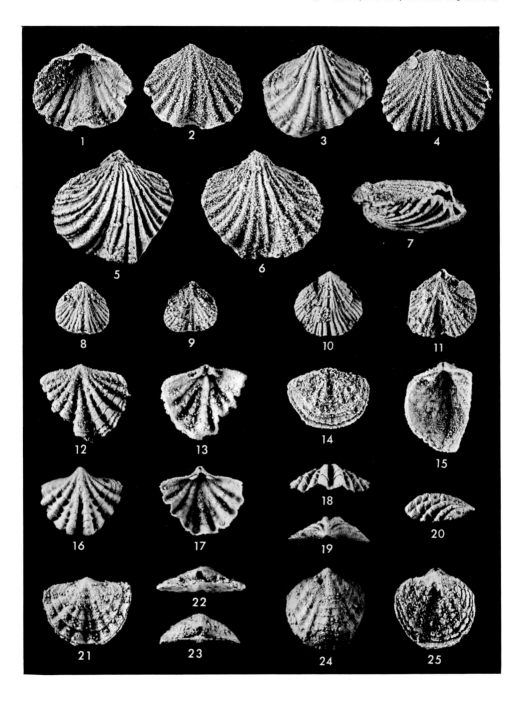

PLATE 26

Figs. 1–18. *Cryptatrypa angusta* n. sp.
    F fauna, Gedinnian, figs. 1–7, 14–18, UCR 5461; figs. 8–13, UCR 5462.
    1–5, Dorsal, ventral, side, anterior, and posterior views × 7, USNM 171534.
    6, 7, Dorsal and ventral views × 7, USNM 171535.
    8, Dorsal view × 4, USNM 171536.
    9, 10, Ventral and dorsal views × 4, USNM 171537.
    11, 12, Dorsal, and ventral views × 4, USNM 171538.
    13, Interior of brachial valve × 7, USNM 171539.
    14, 15, Interior and exterior of brachial valve × 7, USNM 171540.
    16, Interior of articulated valves × 7, USNM 171541.
    17, Interior of articulated valves × 10, USNM 171542.
    18, Interior of brachial valve × 10, USNM 171543.

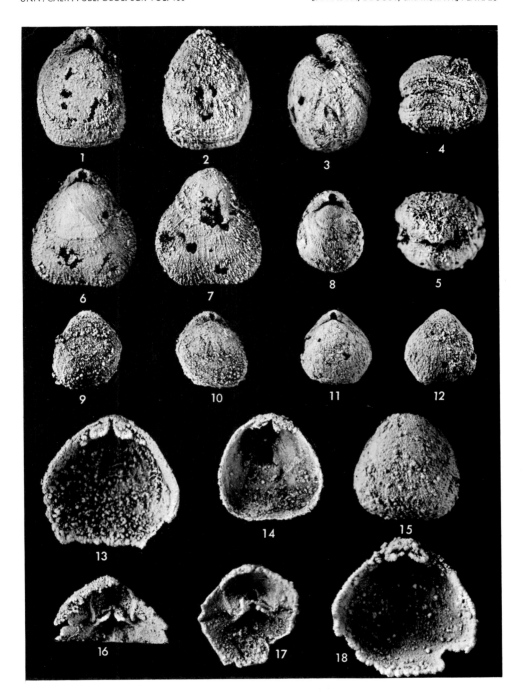

PLATE 27

Figs. 1–7. *Cryptatrypa* sp.
    F fauna, Gedinnian; figs. 1, 2, 5, 6, 7, UCR 5473; figs. 3, 4, UCR 5453.
    1, 2, Exterior and interior views of brachial valve × 3, USNM 171544.
    3, 4, Exterior and interior of brachial valve × 3, USNM 171545.
    5, 6, Ventral and dorsal views × 3, USNM 171546.
    7, Interior of brachial valve × 3, USNM 171547.

Figs. 8–12. *Dubaria* sp.
    F fauna, Gedinnian; figs. 8, 9, UCR 5450; figs. 10, 11, UCR 5473; fig. 12, UCR 5462.
    8, 9, Anterior and ventral views × 2, USNM 171548.
    10, Interior of pedicle valve × 5, USNM 171549.
    11, Interior of both valves articulated × 5, USNM 171550.
    12, Interior of brachial valve × 4, USNM 171551.

Figs. 13–15. *Dubaria megaeroides* Johnson and Boucot
    Pridolian fauna, UCR 5438.
    13–15, Anterior, ventral, and dorsal views × 2, USNM 171552.

Figs. 16–21. *Lissatrypa* sp.
    Basal F fauna, Gedinnian; figs. 16–19, 21, UCR 5441; fig. 20, UCR 5442.
    16, 17, Ventral and dorsal views × 2, USNM 171553.
    18, 19, Ventral and dorsal views × 2, USNM 171554.
    20, Interior of pedicle valve × 3, USNM 171555.
    21, Interior of brachial valve × 3, USNM 171556.

PLATE 28

Figs. 1–9. *Meristella* cf. *wisniowskii* Kozlowski

F fauna, Gedinnian, figs. 1–3, UCR 5461; fig. 4, UCR 5457; figs. 5–9, UCR 5458.

1, Interior of brachial valve × 3, USNM 171557.

2, 3, Side and posterior views of pedicle valve × 1.5, USNM 171558.

4, Interior of pedicle valve × 1.5, USNM 171559.

5–9, Ventral, dorsal, side, posterior and anterior views × 1.5, USNM 171560.

Figs. 10–15. *Nucleospira* sp.

F fauna, Gedinnian, UCR 5458.

10–12, Exterior, interior, and posterior views of pedicle valve × 3, USNM 171561.

13–15, Posterior, exterior, and interior views of brachial valve × 3, USNM 171562.

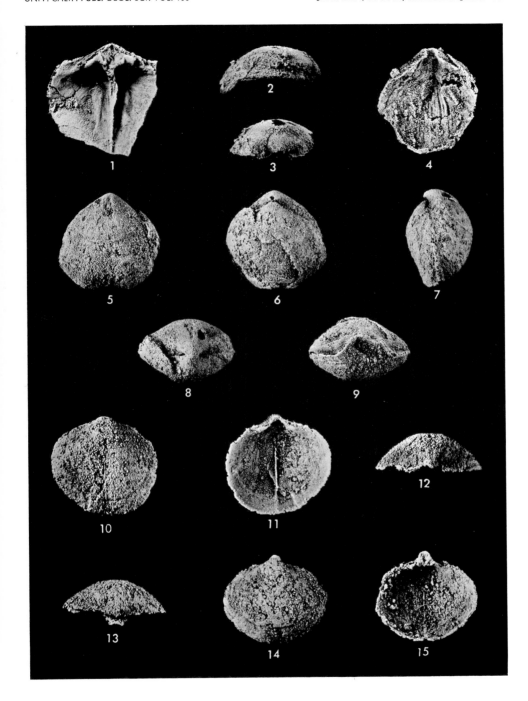

PLATE 29

Figs. 1–17, *Rhynchospirina siemiradzkii* Kozlowski
  F fauna, Gedinnian, figs. 1–7, 10–14, UCR 5459; figs. 8, 9, UCR 5461; figs.
15–17, UCR 5454.
  1–5, Ventral, dorsal, side, posterior, and anterior views × 3, USNM 171563.
  6, 7, Ventral and dorsal views × 4, USNM 171564.
  8, 9, Exterior × 3, and interior × 7 of brachial valve, USNM 171565.
  10, 11, Ventral and dorsal views × 3 USNM 171566.
  12, 13, Ventral and dorsal views × 5, USNM 171567.
  14, Interior of pedicle valve × 3, USNM 171568.
  15–17, Exterior × 2, posterior × 5, and interior × 5 of brachial valve, USNM
    171569.

PLATE 30

Figs. 1–9. *Megakozlowskiella* cf. *M? inflectens* (Barrande)
F fauna, Gedinnian, figs. 1–4, UCR 5469; figs. 5–9, UCR 4449.
1, Exterior of pedicle valve × 5, USNM 171570.
2–4, Interior, posterior, and exterior of brachial valve × 2, USNM 171571.
5–9, Interior, posterior, side, anterior, and exterior of pedicle valve × 3, USNM 171572.

Figs. 10–18. *Undispirifer* cf. *laeviplicatus* (Kozlowski)
F fauna, Gedinnian, figs. 10–15, UCR 5461; fig. 16, UCR 5462; figs. 17, 18, UCR 5460.
10, 11, Side and exterior views of pedicle valve × 2, USNM 171573.
12, 13, Interior and posterior pedicle valve × 3, USNM 171574.
14, 15, Side and interior views of pedicle valve × 2, USNM 171575.
16, Exterior of brachial valve × 3, USNM 171576.
17, 18, Exterior and interior of brachial valve × 3, USNM 171577.

PLATE 31

Figs. 1, 2. *Howellella* sp.
    F fauna, Gedinnian, UCR 5461.
    Exterior and interior of brachial valve × 4, USNM 171578.

Figs. 3–15. *Cyrtina* sp.
    3–5, Exterior, interior and posterior views of brachial valve × 3, USNM 171579.
    6, 7, Interior and exterior of brachial valve × 4, USNM 171580.
    8, 9, Interior and exterior of brachial valve × 4, USNM 171581.
    10–14, Interior, posterior, ventral, side, and anterior views of pedicle valve × 3, USNM 171582.
    15, Posterior view of pedicle valve × 3, USNM 171583.

Figs. 16–19. *Metaplasia?* sp.
    F fauna, Gedinnian, UCR 5462.
    Ventral, side, anterior, and posterior views × 5, USNM 171584.

Figs. 20–28. *Metaplasia lenzi* n. sp.
    F fauna, Gedinnian, figs. 20–25, UCR 5462; figs. 26–28, UCR 5461.
    20, 21, Interior and exterior of brachial valve × 5, USNM 171585.
    22, Interior of brachial valve × 7, USNM 171586.
    23, Interior of brachial valve × 7, USNM 171587.
    24, Interior of brachial valve × 5, USNM 171588.
    25, Exterior of brachial valve × 4, USNM 171589.
    26–28, Side, posterior, and ventral views of pedicle valve × 5, USNM 171590.

ISBN: 0–520–09447–6